Statistical Physics of Polymers

Advanced Texts in Physics

This program of advanced texts covers a broad spectrum of topics which are of current and emerging interest in physics. Each book provides a comprehensive and yet accessible introduction to a field at the forefront of modern research. As such, these texts are intended for senior undergraduate and graduate students at the MS and PhD level; however, research scientists seeking an introduction to particular areas of physics will also benefit from the titles in this collection.

Springer
Berlin
Heidelberg
New York
Hong Kong
London
Milan
Paris
Tokyo

Toshihiro Kawakatsu

Statistical Physics of Polymers

An Introduction

With 129 Figures

Springer

Prof. Toshihiro Kawakatsu
Tohoku University
Department of Physics
Aoba, Aramaki, Aoba-ku
Sendai 980-8578, Japan

Cataloging-in-Publication Data.

Kawakatsu, Toshihiro.
(Kobunshi butsuri no kiso. English)
Statistical physics of polymers: an introduction / Toshihiro Kawakatsu.
p.cm. – (Advanced texts in physics, ISSN 1439-2674)
Includes bibliographical references and index.
ISBN 3-540-43440-2 (acid-free paper)
1. Polymers. 2. Statistical physics. I. Title. II. Series.
QC173.4.P65K3913 2004 530.4'29–dc22 2004042943

Original Japanese edition published by Saiensu-Sha Co., Ltd, Tokyo, 2001

ISSN 1439-2674

ISBN 3-540-43440-2 Springer-Verlag Berlin Heidelberg New York

Springer-Verlag is a part of Springer Science+Business Media

springeronline.com

Typesetting: Data prepared by the author using a Springer TEX macro package
Cover design: *design & production* GmbH, Heidelberg

Printed on acid-free paper SPIN 10838845 57/3141/tr 5 4 3 2 1 0

Preface

This book is an introductory textbook on the statistical mechanics of polymers and complex fluids aimed at senior undergraduate and graduate students and non-specialist researchers who are starting research in this field. Modern statistical mechanics on polymers and complex fluids is based on many fields, such as chemical physics, statistical mechanics, quantum mechanics, stochastic processes, theory of phase transitions, hydrodynamics, rheology, and so on.

This book provides an overview of the basic concepts and methods used in current research on the physics of polymers and complex fluids. Using simple but essential examples, we describe how to derive the physical properties of polymers theoretically, focusing on the structure and dynamics on mesoscopic scales. Here, the term 'mesoscopic scales' means intermediate lengths and time scales between the microscopic atomic scale and the macroscopic scale.

Properties on mesoscopic scales are the central issue of the physics of polymers and complex fluids, because these materials are well characterized by spatiotemporal structures on these scales, where we can extract universal properties that are independent of the microscopic details of the system.

In order to understand universal mesoscopic properties, it is important to eliminate microscopic degrees of freedom irrelevant to these universal properties. Although such an approach has been very successful in describing macroscopic properties of simple liquids, it is not so simple to find out whether this powerful framework can also be applied to complex fluids, because of the coexistence of structures over a wide range of length and time scales.

The complexity of complex fluids usually leads to theoretical models described by nonlinear non-equilibrium equations. In order to extract useful predictions from such theoretical models, we are compelled to rely on analytic approximations or direct numerical calculations. Using polymeric systems as the main target, we show how these numerical calculations and analytical theories can be constructed.

This book is organized as follows:

- Chapter 1 contains a brief overview of various mesoscopic phenomena in complex fluids, together with a brief introduction to the modeling of these phenomena.
- Chapter 2 focuses on polymers, i.e., chain-like macromolecules, as a typical example of complex fluids. We discuss microscopic modeling of the statistical mechanics of such polymer chains. We introduce a molecular model called the Gaussian chain model, which serves as a useful starting point for theoretical studies on the equilibrium and dynamic properties of polymers. In the derivation of this Gaussian chain model, we extensively utilize the so-called coarse-graining procedure, where unimportant microscopic degrees of freedom are eliminated. To demonstrate the efficiency and usefulness of the Gaussian chain model, we compare the theoretical results of this model with recent experimental results concerning biological polymers.
- Chapter 3 further applies the coarse-graining procedure to produce a continuum description of the polymer chain in terms of the density fields and the probability distribution of the shape of the chain. One such theoretical framework is the so-called self-consistent field theory, where many-body problems of dense polymeric liquids and blends are formulated using the mean-field approximation and the probability distributions of the Gaussian chain model. Theoretical as well as numerical analysis using the self-consistent field theory has been widely applied to predict the mesoscopic structures of inhomogeneous polymer systems, such as polymer blends or polymers near surfaces. We will exhibit examples of recent research based on this approach.
- Although the self-consistent field theory is useful and quantitatively reliable, it often requires a heavy numerical calculation. To avoid such a numerical task, we develop another coarse-grained model called the Ginzburg–Landau model in Chap. 4. This model was originally proposed for the critical phenomena associated with second order phase transitions to explain their universal properties. The same technique is applicable to polymer solutions and polymer blends. The model contains several parameters that reflect microscopic information concerning the structure of the polymer chain.
- Viscoelasticity is one of the most important macroscopic properties of dense polymer solutions. It is instructive to see how such macroscopic viscoelastic properties are derived from microscopic molecular models. For this purpose, we discuss reptation theory and its recent advances in Chap. 5.

This book owes much to the kind help of many colleagues. First, I would like to thank Professor Seiji Miyashita (Tokyo University), Mr. Nobuhiko Tajima and Mr. Kosuke Hirase (Saiensu-Sha Co. Ltd.) for their continuous help, encouragement and advice during the preparation of the original

Japanese edition of this book. The publication of the English edition came into being thanks to Dr. Claus Ascheron and Dr. Angela Lahee (Springer-Verlag, Berlin). I also thank Dr. Hiroshi Morita (JST, CREST), Mr. Kenji Furuichi, Mr. Yohei Yamaguchi, Ms. Yuko Mizuta (Nagoya University), Mr. Rihei Endo (Tohoku University), Mr. Yoshikatsu Hayashi (Lund University), Professor Takao Ohta (Kyoto University), Professor Tohru Okuzono (ERATO, JST), and Mr. Takashi Honda (Nippon Zeon Co. Ltd.) for their useful comments on the original Japanese edition.

This book is based on a series of lectures that I gave for graduate students at several universities in Japan. A note on one such lecture at Kyoto University in June 1998 was edited by Dr. Yoshihiro Yamazaki (Waseda University) and Mr. Masataka Imura (Nara Advanced Institute of Science and Technology), and was published in Bussei Kenkyuu (in Japanese). This lecture note was useful in preparing this textbook.

I would also thank authors and publishers for their kind permission to reproduce figures.

Last but not least, discussions with Professor Masao Doi (Nagoya University) and members of the Doi Laboratory and those of the Doi Project sponsored by the Ministry of Economy, Technology and Industry (Japan) are gratefully acknowledged.

Sendai, January 2004 *Toshihiro Kawakatsu*

Contents

1 Introduction 1
1.1 Complex Fluids and Polymers 2
1.1.1 Complex Systems 2
1.1.2 Complex Fluids 3
1.1.3 Mesoscopic Structures in Complex Fluids 7
1.2 Polymers: A Typical Example of Complex Fluids 8
1.2.1 Molecular Structure of Polymers 8
1.2.2 Mesoscopic and Macroscopic Properties of Polymers . . 10
1.3 Modeling the Physical Phenomena of Polymers 11
1.3.1 Modeling Mesoscopic Structures 11
1.3.2 Static Properties 12
1.3.3 Dynamic Properties 14
1.3.4 Physical Properties and Gaussian Chain Statistics 14

2 Gaussian Chain Model and Statistics of Polymers 17
2.1 A Simple Model of a Polymer Chain: The Lattice Model 17
2.1.1 Definition of the Lattice Model of Polymers 17
2.1.2 Ideal Chain Statistics of Lattice Models 25
2.2 Bead–Spring Model of Polymer Chain and Gaussian Chain Statistics 34
2.2.1 Coarse-Graining Procedure and Bead–Spring Model . . 34
2.3 Statistical Mechanical Theory of Equilibrium Conformations of a Gaussian Chain 37
2.3.1 Ideal Chain Statistics 37
2.3.2 Correlation Functions and Scattering Functions of an Ideal Chain 43
2.3.3 Statistical Mechanics of Chains with Interactions and Approximate Theories 48
2.3.4 Statistical Properties of Many-Chain Systems 54
2.4 Dynamical Models of a Polymer Chain Based on a Molecular Description 62
2.4.1 Formulation of Brownian Motion 62
2.4.2 Rouse Model of a Single Polymer Chain in a Solvent . . 66
2.4.3 Hydrodynamic Effects in Dilute Polymer Solutions . . . 73

2.5 Justification of the Gaussian Chain Model from a Microscopic Point of View 80
2.5.1 Full Atomistic Model of Polymer Chains 80
2.5.2 United Atom Model 81
2.6 Statistical Theories and Experiments on Semi-Flexible Chains 84
2.6.1 Worm-Like Chain Model 84
2.6.2 Statistical Properties of a Stretched Worm-Like Chain 86
2.6.3 Experiments on Worm-Like Chains Using Biopolymers 90
2.7 Molecular Simulations of Polymer Dynamics 95
2.7.1 Molecular Simulation Methods 95
2.7.2 Models of Interaction Potentials for Coarse-Grained Chains 96
2.7.3 Examples of Molecular Simulations 96
Exercises 99

3 Mesoscopic Structures and Self-Consistent Field Theory .. 101
3.1 Mesoscopic Phenomena in Polymer Systems 101
3.2 Formulation and Simple Examples of the Self-Consistent Field Theory of Polymers 103
3.2.1 Mean Field Approximation and Self-Consistent Field . 105
3.2.2 Path Integral Formalism for Polymers 106
3.2.3 Classical Approximation for Self-Consistent Field Theory 111
3.3 Numerical Methods for the Self-Consistent Field Theory of Polymers 121
3.3.1 Functionals, Functional Derivatives and Functional Integrals 122
3.3.2 General Expression for the Free Energy 126
3.3.3 Numerical Solutions of Self-Consistent Field Equations 134
3.3.4 Examples of Numerical Simulations Using Self-Consistent Field Theory 142
Exercises 151

4 Ginzburg–Landau Theory 153
4.1 Formulation of the Ginzburg–Landau Theory 153
4.1.1 Ginzburg–Landau Model 153
4.1.2 Expansion of the Free Energy 159
4.1.3 Evaluation of Expansion Coefficients Using the Random Phase Approximation 162
4.2 Applications of the Ginzburg–Landau Theory 172
4.2.1 Phase Diagram of Block Copolymer Melts 172
4.2.2 Extensions to Dynamical Processes 174
Exercises 177

5 Macroscopic Viscoelastic Theory of Polymers 179
5.1 Viscoelastic Properties of Polymeric Liquids 179
5.1.1 Polymers and Viscoelasticity 179
5.1.2 Hydrodynamic Descriptions of Viscoelasticity 186
5.2 Reptation Theory for Linear Polymers 190
5.2.1 Concept of Reptation Theory 190
5.2.2 Stress Relaxation Function 192
5.3 Extensions of Reptation Theory and Nonlinear Viscoelasticity 201
5.3.1 Contour Length Fluctuations 203
5.3.2 Chain Retraction 205
5.3.3 Constraint Release 206
5.3.4 Contribution to Viscoelasticity from Phase-Separated Domains 208
Exercises 210

References 211

Index 213

1 Introduction

Polymers are one of the most popular materials encountered in our daily lives. For example, polymers are the major component of plastics, films, fibers, foods, biomaterials, and so on. Polymers are used extensively in the chemical, electronic, optical, pharmaceutical and medical industries as important components of highly functional materials. To realize this high functionality, the micron-scale or nanoscale structures in these materials usually play an important role. For example, when there are two polymeric components that have different characteristic functions, we can produce a macroscopically uniform material that possesses the functions of both of these components by generating alternating layers of the components on the submicron scale. This technique is made possible by the high flexibility of polymeric materials, which have a large number of internal degrees of freedom within a single molecule. Another important aspect of polymers is their complex physical and mechanical response to externally imposed fields. When a polymeric liquid is deformed by an external field, it exhibits both liquid-like and solid-like dynamical properties, which we call viscoelasticity. The origin of this viscoelasticity is again the complex structure of the polymers.

The above-mentioned complex behavior is not specific to polymers but is shared by a certain class of systems called complex fluids. Recently, these interesting physical aspects of complex fluids have attracted more and more scientists working in the field of statistical physics or condensed matter physics, because of the similarity in concepts and methodology between the physics of complex fluids and solid-state physics.

Because a polymer molecule is a chain-like macromolecule composed of a sequence of various types of monomer units, there can be infinitely many types of polymer depending on the combination and sequence of these monomer units along the chain. When a polymer molecule is dissolved in a solvent, the entire molecule forms a coil structure with a large number of possible folding shapes, because of the high flexibility of the chemical bonds that connect the atoms. Such large numbers of internal degrees of freedom enable us to design polymers with various functions and physical/chemical properties. On the other hand, the complexity associated with many internal degrees of freedom makes it extremely difficult to study polymers theoretically on the microscopic level.

In this chapter, we review the characteristic features of polymers and complex fluids. We will focus on structure and dynamics on length scales much larger than the atomic length scale but still smaller than the macroscopic scale, i.e., the mesoscopic scales. Indeed, many interesting phenomena originate from this mesoscopic structure and dynamics. To treat the large numbers of degrees of freedom of polymers theoretically, statistical mechanics is a useful and efficient tool. In order to prepare for the discussions in the following chapters, we briefly review here the basic concepts and techniques used to study polymers from the point of view of statistical mechanics.

1.1 Complex Fluids and Polymers

1.1.1 Complex Systems

During the past two decades, we have experienced an active and rapid development in research into the so-called complex systems. This has happened in many different fields, such as statistical physics, geophysics, astrophysics, social sciences, etc. As the target of this research, there are roughly two types of system: one is a system composed of large numbers of degrees of freedom that interact mutually, and the other is a system composed of a few degrees of freedom but with highly nonlinear interactions. Due to these features of complex systems, it is difficult to explain their behavior within the traditional theoretical framework developed for simple systems. As mentioned above, polymer molecules have large numbers of internal degrees of freedom. They therefore constitute typical examples of complex systems.

Atmospheric convection is one of the best-known examples of a complex system, where large numbers of degrees of freedom interact with each other in a complex manner. The important variables for this phenomenon are, for example, the distribution of the velocity and temperature of the atmosphere, the geometry of the ground surface, the rotation of the earth, and so on. Nonlinear coupling between these variables creates a rich variety of atmospheric phenomena. The cirrocumuli found in the autumn sky are a visible consequence of such atmospheric convection rolls.

A straightforward approach to this complex problem is numerical analysis of the basic equations of motion. For hydrodynamic systems such as the atmospheric convection mentioned above, we have to solve a set of partial differential equations including the Navier–Stokes equation for hydrodynamic flow and the transport equation for heat due to diffusion and advection. In accordance with the recent rapid development of high-speed supercomputers, such a direct numerical method is becoming more and more popular.

Another approach to the hydrodynamic properties of the atmosphere is to reduce the complex problem to a rather simple problem by eliminating most of the irrelevant degrees of freedom while keeping a few essential degrees of freedom. A well-known example of such an approach is the Lorentz

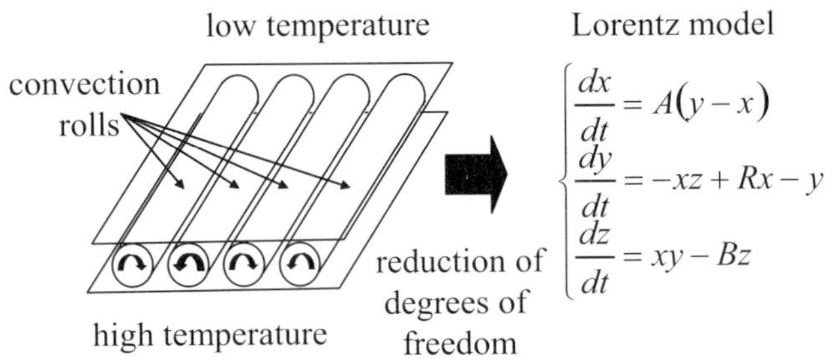

Fig. 1.1. Convection rolls formed in a thin fluid layer heated at the bottom boundary. Lorentz proposed a mathematical model for this convection system using only three variables x, y, and z. Here, A, B, and R are constant parameters accounting for external conditions

model for the dynamics of large-scale atmospheric convection, where the collective motion of convection rolls is described by only three variables (see Fig. 1.1). Detailed numerical simulations of this Lorentz model have shown that an essential unpredictability exists in such a convection phenomenon.[1] The Lorentz model demonstrates that we can understand the essential properties of atmospheric convection without knowing its dynamics on the microscopic scale.

Considering the two theoretical approaches described above, the latter, based on a reduction in the degrees of freedom, is useful for constructing a tractable model of a complex fluid.

1.1.2 Complex Fluids

Besides atmospheric convection systems, there are many complex systems around us. Many materials are categorized as complex fluids, such as rubbers, plastics, surfactants, liquid crystals, biopolymers, and so on. Their characteristic properties are, for example, an extremely large deformation induced by a weak external stress, coexistence of liquid-like behavior and solid-like behavior, strong effects of thermal fluctuations, and so on. All these characteristic features originate from the fact that complex fluids possess mesoscopic structures that are much larger than the atomic size but still much smaller than the macroscopic scales [1–7].

Let us consider a rubber as a typical example of a complex fluid. A rubber is an elastic body with extremely low elasticity. One can stretch the rubber up to a few hundred percent of its original size. Such a response to the external

[1] The three variables in the Lorentz model correspond to the amplitudes of the sinusoidal modulations of the fluid velocities in directions parallel and perpendicular to the fluid layer and the temperature modulation in the direction parallel to the fluid layer.

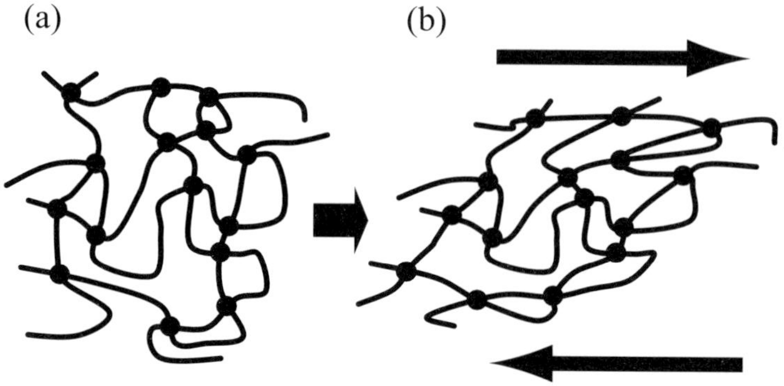

Fig. 1.2. (**a**) The mesoscopic structure of rubber and (**b**) its deformation induced by an external shear stress

force is completely different from that of metals and crystals. The temperature change upon an adiabatic expansion of a rubber is also unusual. While a normal elastic body cools down upon adiabatic expansion, a rubber generates heat when it is stretched quickly. This fact shows that the mechanism of rubber elasticity is very different from that in crystals or metals. A crystal is composed of regularly arranged atoms or molecules and the interaction between the atoms/molecules is the origin of the elasticity. In this sense, the elasticity of crystals is enthalpic. In contrast, there is no regular structure in a rubber. On the microscopic level, the rubber is composed of a network of mutually crosslinked polymer chains, as shown in Fig. 1.2a. When a rubber is stretched, a deformation of its molecular network takes place. Due to such a deformation, the total number of possible spatial arrangements of the network decreases, which leads to a decrease in the entropy. This decrease in the entropy is the origin of rubber elasticity.

Similar situations can be observed quite generally in complex fluids. In this section, we will discuss the origin of the complex behavior of typical complex fluids, focusing on their microscopic internal structure.

Polymer Solutions

Polymers are macromolecules with string-like shape [1]. As we discuss in detail in Chap. 5, dense solutions of polymers with a large molecular weight commonly exhibit viscoelasticity [7, 8], which is a type of behavior intermediate between elasticity and fluidity. Such viscoelasticity originates from the entanglements between polymer chains. Depending on the time scale of the external deformation, a dense polymer solution behaves as an elastic body like rubber or as a viscous fluid such as silicon oil (see Fig. 1.3). Owing to the ease of processing and the variety of molecular structures, polymeric materials have been widely used in many fields of engineering and industry.

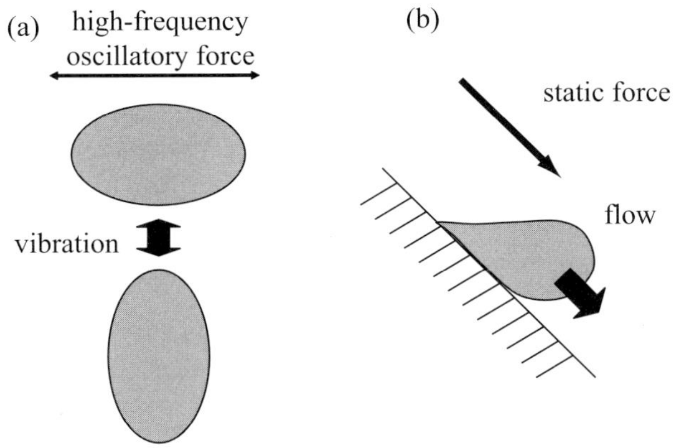

Fig. 1.3. Typical response of a dense polymer solution to external deformation forces. (**a**) Elastic response to a fast deformation and (**b**) fluid-like response to a slow deformation

Gels

Gels are structurally somewhat similar to polymer solutions. The essential difference between gels and polymer solutions is their elastic behavior. Gels are produced by bridging polymer chains with chemical bonds or hydrogen bonds to form a network structure. When an external force deforms a gel, the gel shows a restoring force due to the elastic energy generated by the deformation of the molecular network. The rubber discussed at the beginning of this section is a typical example of a gel. It is easy to deform the flexible network of a gel. The deformation causes a decrease in the entropy of the network, which is equivalent to a deformation elasticity (see Fig. 1.2). To understand the behavior of gels on the molecular level, the statistical mechanical treatment of polymer chains described in Chap. 2 proves very useful.

Surfactants and Emulsions

Examples of complex fluids are not restricted to macromolecular systems. A surfactant solution is one such example, made of low molecular weight components [9, 10]. When a surfactant (detergent) is added to a mixture of oil and water with an appropriate composition, the mixture becomes a uniform opaque liquid. Microscopically, such a solution has a fine structure called an emulsion structure, where a lot of small droplets of oil surrounded by surfactant molecules are immersed in the water phase. Within a certain range of composition and temperature, the characteristic length scale of each emulsion droplet becomes 10–100 nm. Such a solution is thermodynamically stable and is called a microemulsion.

Microemulsions show complex responses to external deformations and external flows, which arise from deformations, fusion and fission of the droplets.

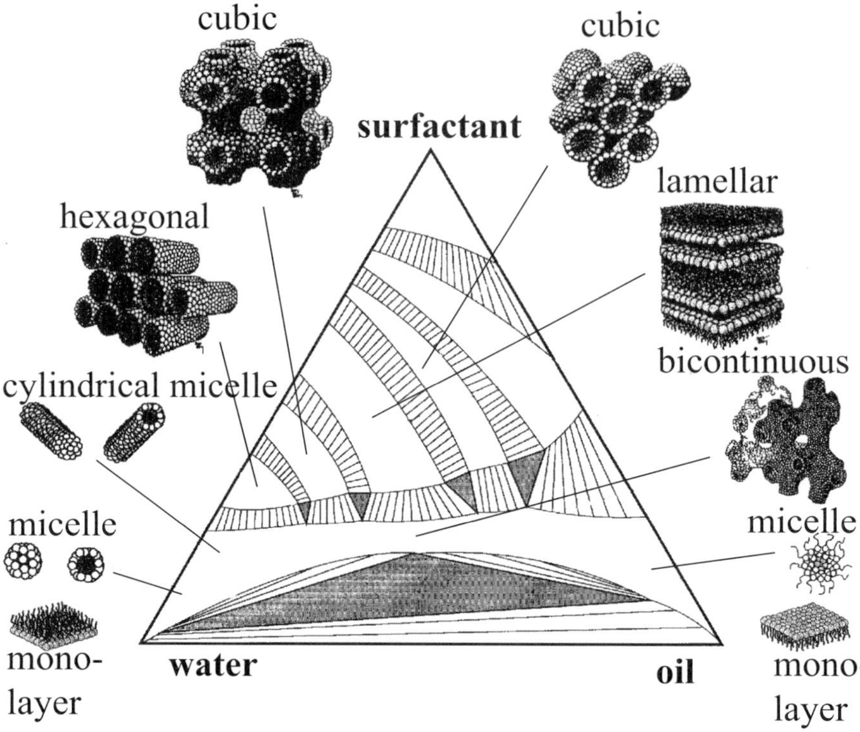

Fig. 1.4. A typical phase diagram of an oil and water mixture containing a surfactant and showing the microscopic structures found in each phase. Reprinted with permission from T. Davis et al., *Physics of Amphiphilic Layers*, ed. by J. Meunier et al. (Springer, 1987)

It is well known that water/oil/surfactant mixtures show a variety of phases with different microstructures depending on the composition of the three components. Figure 1.4 shows a typical phase diagram and microstructures of this mixture at a certain temperature. Each vertex of the triangle corresponds to a pure phase of the water, oil or surfactant, respectively, and any point in the triangle corresponds to a mixture of these three components with a certain composition. Various microstructures are shown in this phase diagram, such as a layered structure, regularly arranged spherical domains, or irregularly arranged membranes. In Chap. 3, we will find that similar domain structures are observed in certain polymer systems with characteristics resembling those of the surfactant.

Liquid Crystals

Another example that resembles the surfactant system is the liquid crystal [11]. As shown in Fig. 1.5a, a liquid crystal molecule has an anisotropic rod-like shape. In a dense solution of such rod-like molecules, various phases

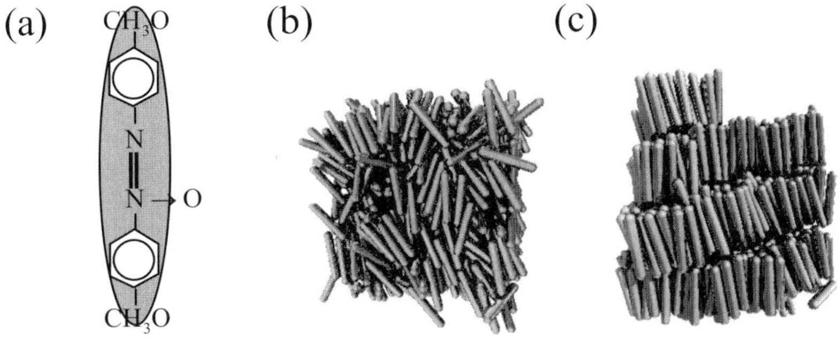

Fig. 1.5. Typical molecular structure of a liquid crystal and examples of the molecular arrangements in dense liquid crystal solutions. (**a**) A typical liquid crystal molecule (*p*-azoxyanisole). (**b**) The nematic phase. (**c**) The smectic phase. Reprinted with permission from S.C. McGrother et al., J. Chem. Phys. **104**, 6755 (1996). (© 1996, American Institute of Physics)

with microscopic structures are formed, such as the nematic phase where the molecules are aligned in a certain direction (Fig. 1.5b), or the smectic phase where a layered structure of molecules is formed (Fig. 1.5c).

Due to the anisotropy of the microstructures in the nematic phase or in the smectic phase, an anisotropy is generated in light transmission or refraction. This optical property of the liquid crystal is used in the liquid crystal display (LCD). The technique for attaching liquid crystalline molecules to a polymer chain by chemical bonds is also used to produce highly viscous or viscoelastic liquid crystalline materials called liquid crystalline polymers.

Biopolymers and Biological Systems

Examples of complex fluids are found not only in engineering materials, but also in biological systems. DNA molecules, actin filaments, and protein molecules are biological polymers. DNA molecules and actin filaments are so large that they are appropriate targets for in situ observation using optical microscopy. In fact, these biological macromolecules have recently been used to test theories of the dynamics of polymers in flow or dense entangled polymers. Examples of fluorescence microscopy images of a DNA molecule and an actin filament are shown in Fig. 1.6. Experimental observations and theoretical consideration of these biopolymers are given in Chap. 2.

1.1.3 Mesoscopic Structures in Complex Fluids

As mentioned above, one basic feature common to complex fluids is the complexity of the internal degrees of freedom, i.e., complex molecular structures (in the case of polymers and gels) or large-scale molecular aggregations (in the case of microemulsions and liquid crystals). These structures exist on intermediate length scales between microscopic atomistic scales and macroscopic

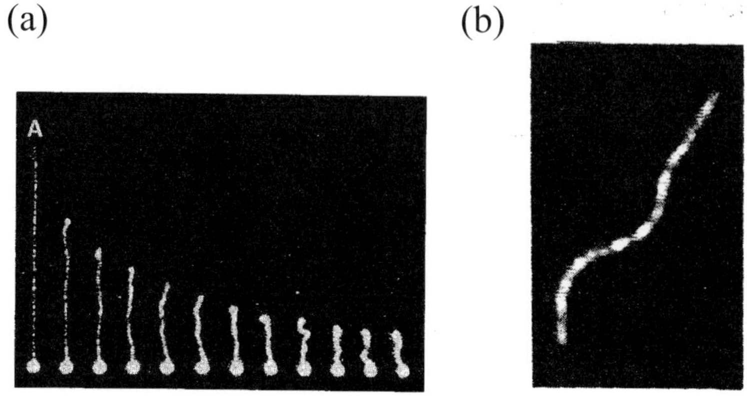

Fig. 1.6. (**a**) Fluorescence microscopy image of a DNA molecule in a uniform flow. One end of the DNA molecule is fixed by laser tweezers. Reprinted with permission from T.T. Perkins et al., Science **264**, 822 (1994). (© 199, AAAS). (**b**) Fluorescent microscopy image of an actin filament. Reprinted with permission from A. Ott et al., Phys. Rev. B **48**, R1642 (1993). (© 1993, American Physical Society)

hydrodynamic scales. These intermediate scales are called mesoscopic scales. Thus the central problem in the physics of complex fluids is to understand phenomena on these mesoscopic scales. The fact that complex fluids possess hierarchical structures including mesoscopic structures means that the statistical mechanical approach, i.e., the reduction of the degrees of freedom to derive simple models, is useful for such complex fluids, just as in the case of hydrodynamic convection discussed at the beginning of this chapter.

In the following chapters, we will study the basic concepts of complex fluids and elementary methods for analyzing these systems theoretically. We will focus on structure and dynamics on mesoscopic scales. Although we will mainly discuss polymers as a typical example of complex fluids, the same concepts and techniques are equally applicable to all complex fluids.

1.2 Polymers: A Typical Example of Complex Fluids

1.2.1 Molecular Structure of Polymers

Plastic films used for wrapping goods are made of a polymer called polyethylene, and the plastic bodies of domestic electrical appliances such as personal computers or audio systems are made of another polymer called high-impact polystyrene. Many components of foods and indeed of our own bodies are also polymers.

Although polymers are so common in our daily life, our current understanding of the fundamental character of these molecules exists mainly on a phenomenological level. This is due to the existence of complex structures on mesoscopic scales and to the coexistence of many spatiotemporal phenomena over a wide range of scales. Let us consider this problem using polystyrene as an example.

As shown in Fig. 1.7a, a polystyrene molecule is composed of a repeated sequence of the unit $-(CH_2-CH(C_6H_5))-$ connected by chemical bonds. In order to synthesize this polystyrene molecule, the styrene molecules $C_6H_5CH{=}CH_2$ shown in Fig. 1.7b must be repeatedly connected by chemical bonds. Such a procedure for producing a polymer chain is called a polymerization reaction. The low molecular weight materials used to produce polymers are called monomers. On the other hand, the atomic group repeated along the polymer chain is called a repeating unit. For example, in the case of polystyrene shown in Fig. 1.7, the styrene molecule corresponds to the monomer, and the atomic group $-(CH_2-CH(C_6H_5))-$ corresponds to the repeating unit.

In the example shown in Fig. 1.7, the linear sequence of carbon atoms is called the backbone. The chemical or physicochemical properties of the polymer originate from the sequence of atoms along this backbone and also from the atomic groups that are connected to this backbone (the benzene groups in the present example), and from the side chains (hydrocarbon chains similar to the backbone). For example, the color and mechanical strength of a plastic material fade when it is subjected to ultraviolet light. Such phenomena result from the photoactivation and breaking of chemical bonds in the polymer chain induced by the absorption of photons. The polymerization or the dissociation reaction of chains takes place at certain atomic groups called functional groups. The ionization of a polymer also takes place at the ionizable atomic groups. These chemical properties of polymers depend strongly on the microscopic atomic species that compose the polymer chain. These properties can therefore change dramatically from one polymer material to another.

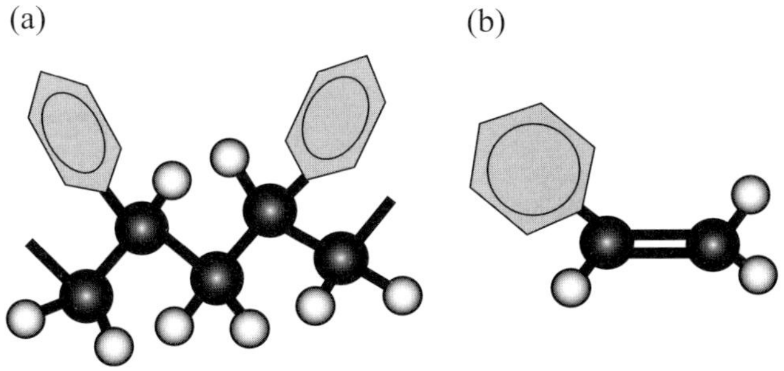

Fig. 1.7. Chemical structures of (**a**) a polystyrene molecule and (**b**) a styrene monomer. *Large* and *small spheres* indicate carbon atoms and hydrogen atoms, respectively. Benzene groups are shown by *hexagons*

1.2.2 Mesoscopic and Macroscopic Properties of Polymers

Despite the various chemical properties described in the last section, polymers show certain universal properties that do not depend on their microscopic details. Mechanical properties and flow properties are examples of such universal properties, and are referred to as the physical properties.

The physical properties of a polymer are mainly determined by its structure on larger scales than the atomic scale. On such larger scales, the polymer chain can be regarded as a string-like object, as shown in Fig. 1.8, where the local chemical structures, the small side groups and even the short side chains are no longer visible. Therefore, we can expect the universal physical properties of polymers to originate from the flexible string-like structure formed by the backbone.

Figure 1.9 shows two typical examples of physical properties of polymeric systems. The first example is the dependence of the diffusion constant on the molecular weight (Fig. 1.9a). In dense polymer systems, the constituent polymer chains are mutually entangled and the diffusion of a chain is constrained by the network formed by the other chains. This diffusion constant is known to be a decreasing function of the molecular weight with an inverse 2–2.3 power. Further, this power law is independent of the type of chain, as long as the chain is long enough and has a linear structure.

The second example is the phase separation behavior of a polymer blend, i.e., a mixture of different types of polymer (Fig. 1.9b). In a multicomponent polymer blend, repulsive interaction between different types of polymer induces phase separation, where the system is divided into many domains composed of a uniform single phase. Many polymer blends show qualitatively similar phase separation behavior, such as domain structures and their growth rate. (Readers should refer to Chaps. 2 and 3.) In this book, we focus mainly on these universal properties of polymer systems, based on the statistical properties of string-like molecules.

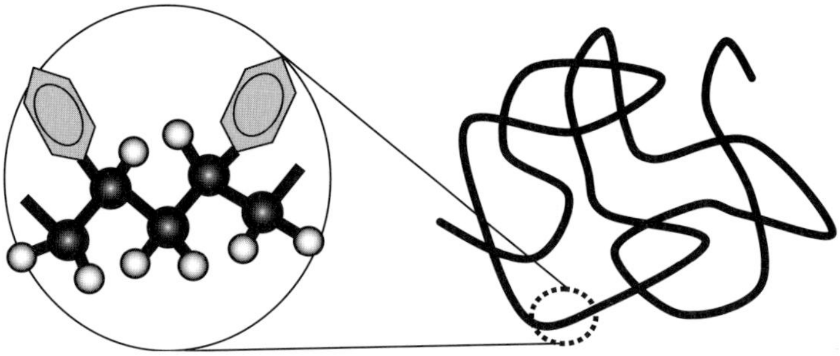

Fig. 1.8. Local structure of a chain and spatial structure of the whole chain. As a whole, a chain can be regarded as a flexible, string-like object

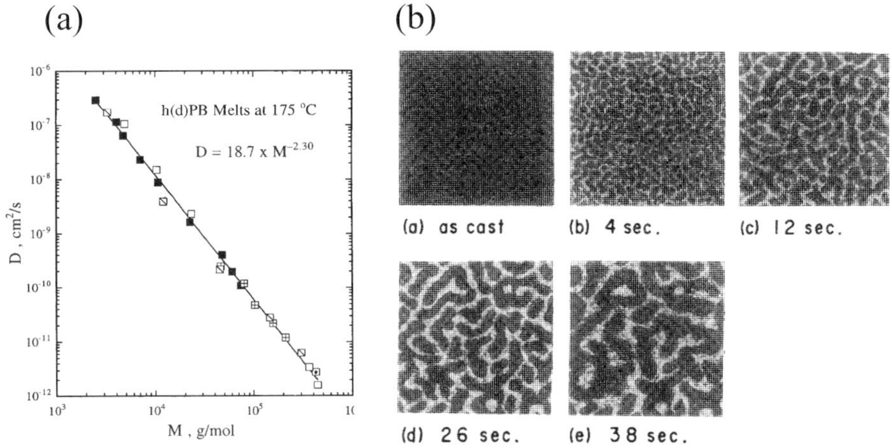

Fig. 1.9. Examples of physical properties of polymers. (**a**) Molecular weight dependence of the diffusion constant of a dense polybutadiene solution. Reprinted with permission from T.P. Lodge, Phys. Rev. Lett. **83**, 3218 (1999). (© 1999, American Physical Society) (**b**) Domain structures caused by a phase separation. Reprinted with permission from A. Nakai et al., Macromolecules 3010 (1986). (© 1986, American Chemical Society)

1.3 Modeling the Physical Phenomena of Polymers

1.3.1 Modeling Mesoscopic Structures

As mentioned in the last section, properties of the polymer chain can be categorized into two classes: chemical properties and physical properties. The former mainly originate from the microscopic structure of the polymer and the latter from the mesoscopic structure.

The most important scales in polymer systems are the mesoscopic scales between the atomic scale and the macroscopic scale. As the typical length scale of atoms and atomic groups is around 0.1–1 nm and the typical length scale of our everyday lives is around 10^{-2}–1 m, the above-mentioned mesoscopic scale actually covers an extremely large regime, extending over nine decades. This suggests that a variety of phenomena must coexist even in the mesoscopic regime. We should also note that microscopic and mesoscopic phenomena are in general coupled together. A quantitative description of physical properties thus requires a unified understanding of various properties on different length and time scales.

1.3.2 Static Properties

Typical examples of static properties of polymers are as follows:

- The constituent units of the chain (monomers or segments).[2]
- The branching structure of the backbone.
- The arrangement of the constituent units along the chain and the total number of these units (molecular weight).
- The interactions between the units.
- The 3-dimensional structure of the chain (chain conformation).
- The molecular aggregation and domain structures generated in the phase separation process.

These examples are illustrated schematically in Fig. 1.10. The branching structure and the arrangement of the monomers along the chain are generally called the configuration, or primary structure (Figs. 1.10a and b). Although the terminology 'configuration' is usually used to distinguish different optical isomers due to the existence of asymmetric carbon atoms on the backbone of the polymer chain, we include the branching structures and the sequence of monomers along the backbone as well. Once a polymer chain is synthesized, its configuration is preserved as long as the chemical bonds are not broken.

As shown in Fig. 1.11a, the branching structures can be classified as linear-type, comb-type, star-type, and so on. On the other hand, as depicted in Fig. 1.11b, the configuration of a linear chain is classified into two categories according to the arrangements of monomers along the chain. One is the class of homopolymers and the other comprises the copolymers. A homopolymer is composed of a single type of monomer, while a copolymer is composed of two or more different types of monomer. The copolymers are further classified into block copolymers, made by connecting different types of homopolymer by chemical bonds, random copolymers, composed of several types of monomer arranged randomly along the chain, and so on. The branching structure and the arrangement of monomers determine the configuration of the polymer.

In polymer solutions, the polymer chains are folded due to the interaction between monomers and solvent molecules (Fig. 1.10c) and also due to the entropy effect. When solvent molecules have an affinity with the monomers, they permeate the folded polymer and swell it. In such a case, the polymer chain forms a randomly folded swollen coil, which is called the random coil state. On the other hand, if the solvent molecules dislike the monomers, they are excluded from the polymer, and the polymer chain is compactly folded without the solvent. Such a state is called the globular state. In each of these states, the polymer chain can assume many different 3-dimensional folded shapes (Fig. 1.10d). Such 3-dimensional shapes of the polymer chain are called conformations, in contrast with the term 'configuration'.

---

[2] A segment is the smallest unit of the polymer chain used in its statistical mechanical description. A segment generally differs from the monomer or the repeating unit. A precise definition of the segment will be given in Sects. 2.1.1 and 2.2.

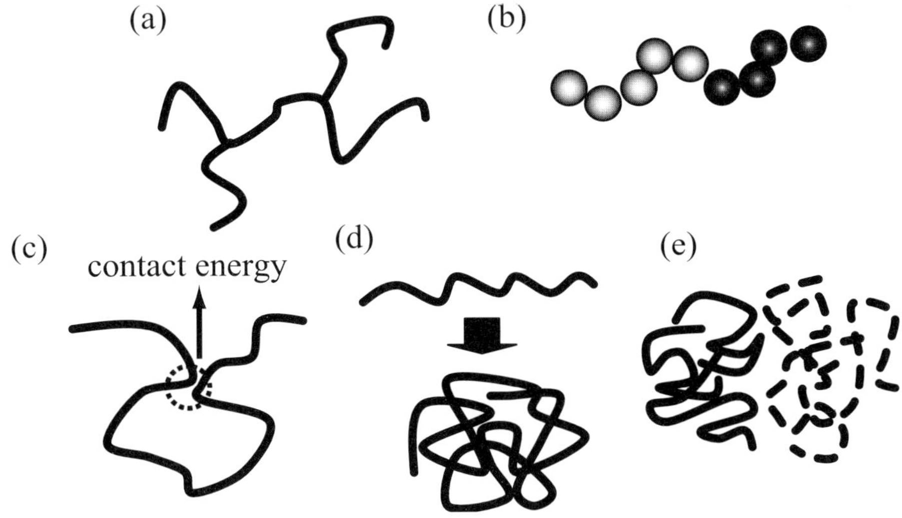

Fig. 1.10. Important factors determining the static features of polymer systems. (**a**) Branching structure of the backbone. (**b**) Arrangement of different types of unit along the chain. (**c**) Interaction between units. (**d**) Three-dimensional folding structure of the chain. (**e**) Aggregation of chains

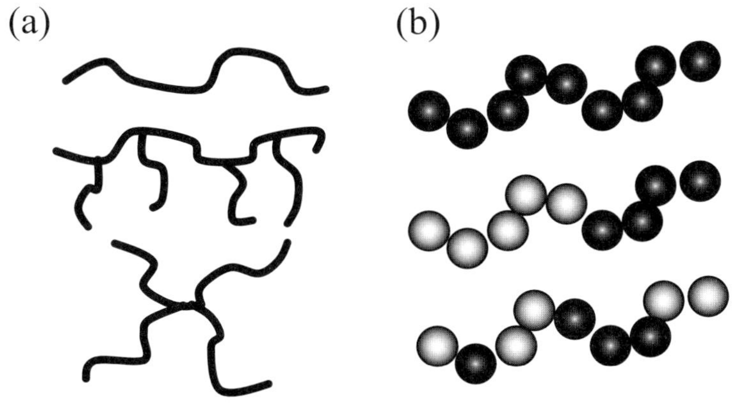

Fig. 1.11. Examples of configurations of a polymer chain. Several types of branching structure and arrangements of monomers are shown. (**a**) Linear, comb, and star structures. (**b**) Homopolymer, block copolymer, and random copolymer

When several types of polymer are mixed, large-scale structures are often formed, e.g., a phase-separated domain structure (Fig. 1.10e), in which the system is divided into many domains with different compositions.

1.3.3 Dynamic Properties

Typical examples of dynamic properties of polymers are as follows:

- Motion of an individual unit of the chain.
- Motion of an individual chain as a whole.
- Cooperative dynamics of many chains (diffusion, viscoelasticity, etc.).
- Growth, shrinkage, fusion and fission of the domains.

Examples of the dynamical properties are shown schematically in Fig. 1.12. Due to the connectivity of the chain, the motion of an individual unit of the chain (Fig. 1.12a) induces a deformation of the chain as a whole (Fig. 1.12b). In a dense polymer mixture, diffusion of one type of polymer into another (interdiffusion) also takes place, and this leads to the formation of mesoscopic domain structures (Fig. 1.12c).

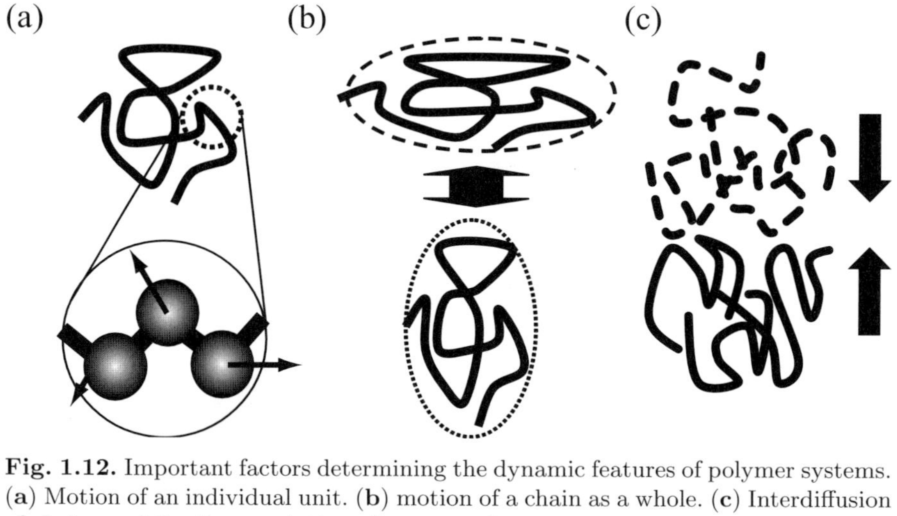

Fig. 1.12. Important factors determining the dynamic features of polymer systems. (**a**) Motion of an individual unit. (**b**) motion of a chain as a whole. (**c**) Interdiffusion of chains and the time evolution of mesoscopic structures

1.3.4 Physical Properties and Gaussian Chain Statistics

The physical properties summarized in the last section can in principle be derived from the microscopic chemical properties on atomic length scales. However, even if we do not have such knowledge of the microscopic chemical details, we can correctly predict the mesoscopic or macroscopic physical properties, at least on a qualitative level.

The theoretical methods that give the correct qualitative mesoscopic behavior of polymer systems are mainly based on the Gaussian chain model. In this model, as shown in Fig. 1.13, the polymer chain is conceived as a sequence of spherical beads connected by freely-rotating linear springs. This Gaussian chain model is very widely used in calculating the chain conformations and motion of chains theoretically. The model is useful not only in describing the conformation and motion of individual chains, but also in describing the mesoscopic collective motion of many chain systems. The latter includes the viscoelastic properties of dense entangled polymer systems, the dynamics of phase separations on mesoscopic scales, and so on. It is one of the main purposes of this book to describe how the various phenomena on different scales are mutually related using this Gaussian chain model.

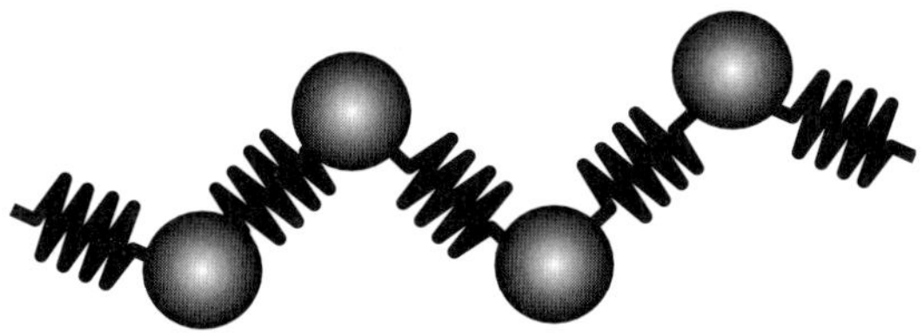

Fig. 1.13. Model of a polymer chain using a sequence of beads connected by springs

2 Gaussian Chain Model and Statistics of Polymers

In this chapter, we will discuss the statistical theory of polymers based on molecular descriptions, where we model the polymer chain as a sequence of spherical particles connected by rigid rods or springs.

Even in such molecular models, we must adopt the idea of coarse-graining to reduce the degrees of freedom. For this reason, we usually assume that the size of the unit of the molecular model is much larger than the atomic scale. One such example is the Gaussian chain model, which provides a basis for the statistical mechanics of polymers. We will learn that this Gaussian chain model can describe the physical properties of many different polymeric systems on longer length and time scales. The Gaussian chain model is therefore a useful starting point for studying the physical properties of polymeric systems.

To outline the theory of Gaussian chain statistics, we first discuss the simplest model exhibiting Gaussian chain behavior, i.e., the lattice model of polymer chains. Next, we introduce dynamics into the Gaussian chain model, and apply it to dynamic problems such as the deformation of a chain in a flow. In order to establish a connection between realistic systems and the Gaussian chain model, we explain how to derive the Gaussian chain model from microscopic atomistic models. We also consider several extensions of the Gaussian chain model so that it can describe biological polymers. We show how the model reproduces the essential behavior of biological polymers observed in recent experiments on DNA molecules and actin filaments. Finally, to stress the importance of numerical methods in polymer science, we present several examples of computer simulations using molecular models of polymers.

2.1 A Simple Model of a Polymer Chain: The Lattice Model

2.1.1 Definition of the Lattice Model of Polymers

Lattice Model of a Polymer Chain

As discussed in Chap. 1, in order to understand the mesoscopic physical properties of polymers, it is important to focus on the flexible string-like

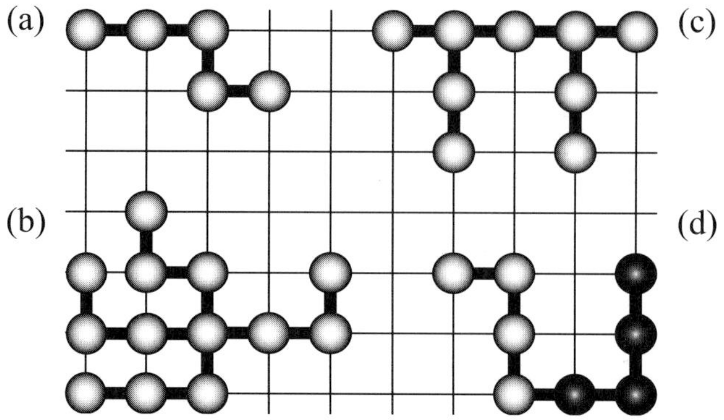

Fig. 2.1. Lattice models of polymer chains. Several branching structures (chain topologies) on a lattice are shown. (**a**) Linear polymer. (**b**) Star polymer. (**c**) Comb polymer. (**d**) Copolymer

structure of the backbone of the polymer chain. One of the simplest models for this purpose is the lattice model illustrated in Fig. 2.1.[1] Although the lattice model is so simple, we will show in Sect. 2.5 that the behavior of the lattice model is consistent with that of realistic atomistic models of polymers on a broad length scale. It is thus instructive to encounter the analytical treatment of the lattice model here as an introduction.

First, we introduce a spatial lattice structure. In order to keep the discussion simple, we use a two-dimensional square lattice and a three-dimensional cubic lattice. As shown in Fig. 2.2, we assume that the lattice has lattice constant b_0 and that the number of nearest lattice sites is z. ($z = 4$ for the two-dimensional square lattice and $z = 6$ for the three-dimensional cubic lattice.) Note that the following discussions do not depend on the details of the lattice structure or on the number of nearest neighbor lattice sites z.

A lattice model of a polymer chain is described by a set of particles placed sequentially on the spatial lattice structure. The constituent particles forming the lattice polymer chain are called segments (see Fig. 2.2). The string-like structure of a polymer chain is modeled by bonds that connect successive segments placed on nearest neighbor lattice sites. It should be noted that the segment is a different concept from the monomer or the repeating unit, which is the chemical unit of the polymer chain.[2] In general, a segment cor-

[1] Although there are no lattice structures in real polymer systems, statistical properties of both the lattice models and realistic atomistic models reduce to the same Gaussian chain statistics as long as we are interested in the large scale phenomena (see Sect. 2.1.2). Such a drastic simplification of the target system is often important in discussing the essential and universal features.

[2] As explained in Sect. 1, the monomer is different from the repeating unit. However, we will not distinguish these two terminologies in this book.

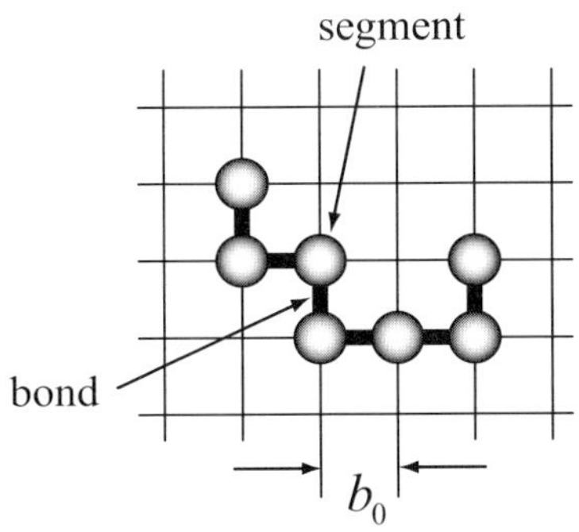

Fig. 2.2. Definitions of the segment and the bond in a lattice model of polymers

responds to a set of monomers (usually a few monomers, but sometimes tens or hundreds of monomers), and is the minimum element of the statistical description of the polymer chain. The rigorous definition of the segment will be given in Sect. 2.2.

The length of a polymer chain is measured by the total number of segments, or the total number of repeating units, or the molecular weight, all of which are proportional to the chain length.

Configuration and Conformation of a Polymer Chain

In a lattice model, each bond connecting the segments is assumed to have a length b_0 which is the same as the lattice spacing. We also assume that the bonds will be neither disconnected nor reconnected under normal conditions. Therefore, the branching structure of the polymer chain is preserved. Such branching structure is sometimes called the chain topology. Figure 2.1 shows several examples of these chain topologies using a lattice model, such as a linear polymer (Fig. 2.1a), a star polymer (Fig. 2.1b), and a comb polymer (Fig. 2.1c). By connecting different types of segments, one can also synthesize a copolymer, as illustrated in Fig. 2.1d. Thus the lattice model of a polymer can correctly describe the topology of the chain and the arrangement of segments along the chain, i.e., the configuration.

Next we discuss how to arrange a polymer chain on the lattice. Due to its flexibility, a polymer chain with a certain configuration can exhibit a very large number of possible 3-dimensional folding shapes (see Fig. 2.3). This degree of freedom is called the conformation. A polymer chain thus has two distinct degrees of freedom: one is the configuration that specifies the primary structure of the chain, and the other is the conformation that specifies the 3-dimensional shape of the chain.

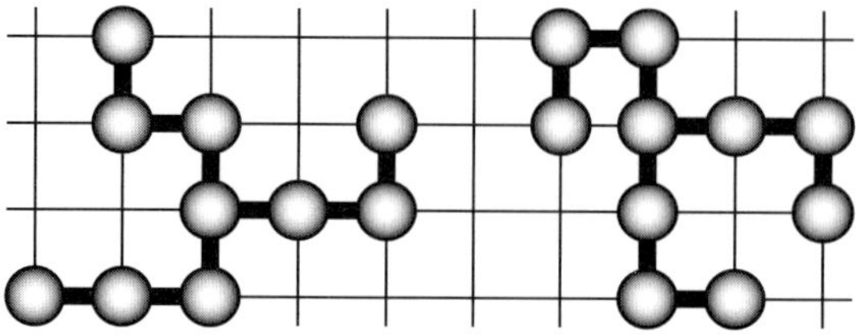

Fig. 2.3. Examples of different conformations of polymer chains with the same configuration

Polymer Conformation and Random Walk Model

One of the important problems in the statistical mechanics of polymers is to calculate the statistical probability distribution of the equilibrium chain conformations. (This corresponds to an evaluation of the entropy of the chain conformation.) Let us evaluate the total number of possible conformations of a single linear polymer chain as an example.

As shown in Fig. 2.4, we consider a linear polymer chain composed of $N+1$ segments, numbering the segments $0, 1, 2, \ldots, N$ from one end to the other. Similarly, we number the N bonds $0, 1, 2, \ldots, N-1$.

If we denote the position vector of the ith segment by $\boldsymbol{r}_i$, the ith bond vector is defined by

$$\boldsymbol{u}_i \equiv \boldsymbol{r}_{i+1} - \boldsymbol{r}_i \quad (i = 0, 1, 2, \ldots, N-1) . \tag{2.1}$$

Then a chain conformation is described by the set of $(N+1)$ segment positions $\{\boldsymbol{r}_i\} \equiv \{\boldsymbol{r}_0, \ldots, \boldsymbol{r}_N\}$ or by the set of N bond vectors $\{\boldsymbol{u}_i\} \equiv \{\boldsymbol{u}_0, \ldots, \boldsymbol{u}_{N-1}\}$. The difference in the number of degrees of freedom between the position vectors and the bond vectors comes from the translational degrees of freedom of the center of mass of the chain, which does not contribute to the chain conformation.

The statistical properties of the chain conformation depend on the type of externally imposed constraints. In the simplest model, the bond vectors $\{\boldsymbol{u}_i\}$

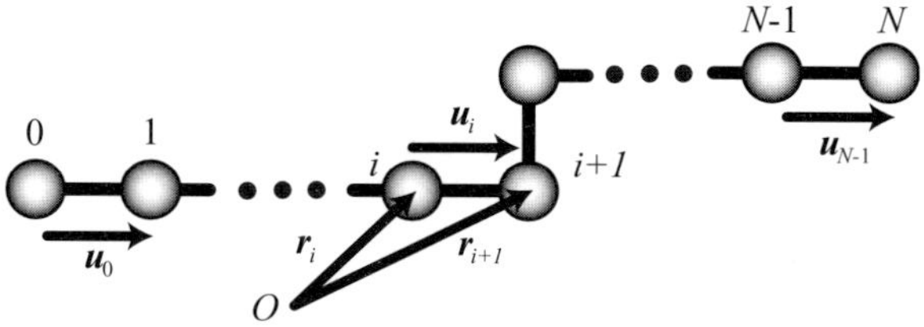

Fig. 2.4. Numbering of the segments from one end of the chain and definition of a bond vector as the vector connecting adjacent segments

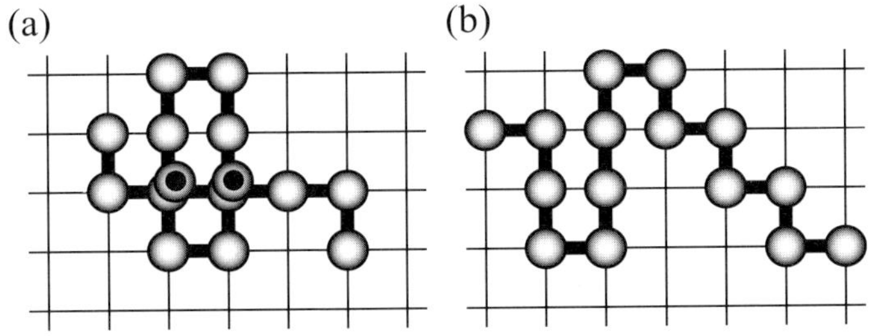

Fig. 2.5. Simple rules for the arrangement of segments on a lattice. (**a**) The ideal chain model where two segments can occupy the same lattice site (a random walk), and (**b**) the excluded volume chain model where two segments cannot occupy the same lattice site (a self-avoiding walk). In case (**a**), the segments marked with *small black dots* are overlapping segments

are assumed to be statistically independent of each other. Thus, the position of the $(i+1)$th segment is chosen at random from the z nearest neighbor lattice sites to the site occupied by the ith segment. In this case, as shown in Fig. 2.5a, we allow two or more segments to occupy the same lattice site.

Such a chain model is called the ideal chain model. In the ideal chain model, the chain conformation can be identified with the path of a random walk of a Brownian particle on a lattice, when we regard the segment index i as time (see the box entitled *Statistical Properties of Random Walks*). Thus the chain conformation usually adopts a random coil structure. The correspondence between the polymer chain conformation and the diffusion process of a Brownian particle will be used in the self-consistent field theory described in Chap. 3.

We can also consider another model of a polymer chain with the constraint that two segments are not allowed to occupy the same lattice site (see Fig. 2.5b). Such a constraint expresses the so-called excluded volume effect of the segments, and this chain model is called an excluded volume chain model or a realistic chain model. The conformation of an excluded volume chain corresponds to a random walk of a particle that will never revisit a position it has already visited before. Such a constrained random walk is called a self-avoiding random walk. As the segments of the excluded volume chain do not overlap each other, the radius of the coil of the chain becomes larger than that of the ideal chain with the same length. Such an expansion of the chain is referred to as swelling. It originates from the effect of the solvent that is not explicitly described in the lattice model. As the polymer chain shown in Fig. 2.5 is not floating in the vacuum but is floating in a solvent, a vacant lattice point is actually occupied by a solvent particle. Therefore, the expansion of the coil of the chain means that the solvent molecules can permeate into the coil.

Statistical Properties of Random Walks

The lattice random walk model is defined by a particle that hops randomly from each lattice site to one of its nearest neighbor lattice sites. On the other hand, the random flight model is defined by a particle that repeatedly and randomly selects a direction and moves through a fixed distance in that direction.

When viewed from a much larger length scale compared with the single hopping distance, the random paths of these two models are statistically identical. They are everywhere undifferentiable, and any small portion of these curves has the same statistical properties as the original curves. Such curves are called fractal curves. In fact, if the space dimension is greater than or equal to 2, the large-scale limit of these random walks is a fractal curve with fractal dimension 2 [12]. After a sufficiently long time, the size of the region covered by the random walk is proportional to the square of the length of the curve (see also the box entitled *Ideal Chains and Fractal Structures* in Sect. 2.3.1.)

Let us consider a random walk of a particle in a d-dimensional space, and denote the position of the particle at time t by $\boldsymbol{r}(t)$. For such a random walk, the mean square displacement $\langle|\boldsymbol{r}(t)-\boldsymbol{r}(0)|^2\rangle$ is known to be proportional to the time t. Then the quantity D defined by

$$D \equiv \lim_{t\to\infty} \frac{1}{2dt}\langle|\boldsymbol{r}(t)-\boldsymbol{r}(0)|^2\rangle \tag{2.2}$$

is called the diffusion constant, which measures the rate of diffusive displacement of the particle.

Excluded Volume Parameter

Whether the solvent permeates inside the coil or not is determined by the competition between enthalpic and entropic effects. At higher temperatures, the segments and solvent molecules tend to mix due to mixing entropy effects, and the polymer chain is therefore swollen. On the other hand, at lower temperatures, if the interaction between the segment and the solvent is repulsive, they tend to separate, and the solvent is excluded from the polymer coil. The excluded volume parameter v expresses this tendency for the chain to swell. It is defined as the spatial volume of the region where a segment excludes the other segments. In the case of the lattice model of the excluded volume chain shown in Fig 2.5b, the excluded volume parameter v equals the volume per lattice site b_0^3. On the other hand, in the ideal chain case, $v = 0$ because there is no excluded volume interaction.

In general, the pair interaction between non-ionic atoms is composed of two different contributions, i.e., a short-range repulsion and a long-range

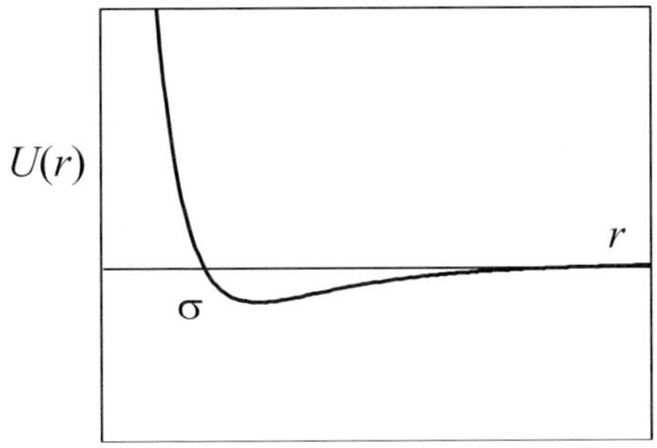

Fig. 2.6. Typical functional form of the interaction potential between neutral atoms

attraction. When the interatomic distance becomes smaller, the electrons of these two atoms start to overlap. A strong repulsion then takes place due to the Pauli exclusion principle. This repulsive part of the interatomic interaction is called the hard core. When the interatomic distance is larger, an attractive interaction takes place between the electric dipoles on the two atoms: one of the dipoles is generated by an instantaneous deformation of the electron cloud and the other is induced by this instantaneous dipole. This attractive interaction originating from the instantaneous dipoles is called the van der Waals interaction.

Combining the above-mentioned repulsive and attractive interactions, the overall interatomic interaction potential between two non-ionic atoms takes the functional form shown in Fig. 2.6. One of the model potentials that mimics such behavior is the following Lennard–Jones potential (usually abbreviated to the LJ potential):

$$U(r) = 4\epsilon\left[\left(\frac{\sigma}{r}\right)^{12} - \left(\frac{\sigma}{r}\right)^{6}\right] . \tag{2.3}$$

Here, r is the interatomic distance, and σ and ϵ are measures of the atomic diameter and interaction energy, respectively.

Although the interaction potential between two segments is similar to the LJ potential, the existence of the solvent molecules makes the situation more complex. The excluded volume parameter v does not correspond to the direct interaction between segments but to the effective interaction that is obtained after the degrees of freedom of the surrounding solvent molecules are replaced by their thermal equilibrium distribution. Such a procedure is called elimination of the degrees of freedom. The effective interaction thus obtained includes, besides the direct interaction between the segments, the segment–solvent interactions, the solvent–solvent interactions, and the entropy effects associated with the arrangements of the solvent molecules as well. The excluded volume parameter thus corresponds to the spatial volume

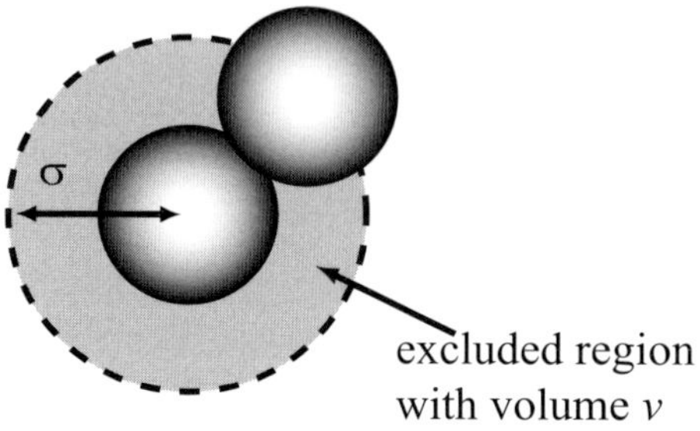

Fig. 2.7. The physical meaning of the excluded volume parameter v when the interaction potential can be approximated by the hard sphere potential

from which a segment excludes other segments under the influence of the solvent molecules.

In terms of the effective interaction potential between the segments $U_{\text{eff}}(r)$, the excluded volume parameter is obtained as the coefficient of the second order term in the cluster expansion of the partition function, and is given by [13]

$$v = \frac{1}{2} \int \mathrm{d}\boldsymbol{r} \left[1 - \mathrm{e}^{-\beta U_{\text{eff}}(r)} \right] , \tag{2.4}$$

where $\beta = 1/k_{\text{B}}T$. This expression shows that the repulsive and attractive parts of the effective interaction give positive and negative contributions to the excluded volume parameter.[3] This excluded volume parameter v corresponds to the second virial coefficient of a dilute gas. If the segment is a hard sphere with diameter b_0, the interaction potential $U_{\text{eff}}(r)$ is infinite for $r < b_0$ and zero for $r \geq b_0$. Then, using (2.4), we obtain $v = (2\pi/3)b_0^3$. This corresponds to the volume exclusively occupied by a single hard sphere (see Fig. 2.7).

Coil–Globule Transition

The conformation of a polymer chain changes its behavior drastically at $v = 0$. When $v > 0$, the interaction between segments is repulsive. In such a case, the chain is swollen and adopts a coil state. On the other hand, if $v < 0$, the segments attract each other and form a spherical aggregate where the segments are densely packed. Such a spherical aggregate is called a globular state. The boundary between the coil state and the globular state is located at $v = 0$, where the chain behaves as an ideal chain.

For each of the three cases $v > 0$, $v = 0$ and $v < 0$, the solvent is called a good solvent, θ-solvent, and poor solvent, respectively. The excluded

[3] An expression for the excluded volume parameter v based on the lattice model will be given in (2.71).

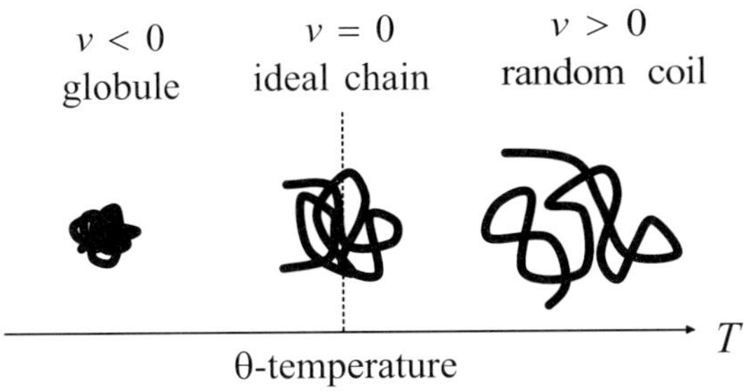

Fig. 2.8. Change in the chain conformations above and below the θ-temperature

volume parameter consists of both the enthalpic and entropic contributions, the relative balance of which changes with the temperature. Hence the quality of the solvent is a function of the temperature. There is a special temperature called the θ-temperature at which the excluded volume interaction vanishes, i.e., $v = 0$. When the temperature decreases across the θ-temperature, the chain conformation changes from a coil state to a globular state. Such a transition is called the coil–globule transition. Figure 2.8 shows a schematic picture of the coil–globule transition.

2.1.2 Ideal Chain Statistics of Lattice Models

Bond Orientational Correlations of an Ideal Chain

We now begin to consider the mathematical treatment of the conformation of a single ideal chain using the lattice model. (As the segments of an ideal chain do not interact with each other, the statistical properties of a set of ideal chains reduce to those of a single ideal chain.) A particular chain conformation is specified by a set of positions of segments $\{\boldsymbol{r}_i\} \equiv \{\boldsymbol{r}_0, \boldsymbol{r}_1, \ldots, \boldsymbol{r}_N\}$, or by a set of bond vectors $\{\boldsymbol{u}_i\} \equiv \{\boldsymbol{u}_0, \boldsymbol{u}_1, \ldots, \boldsymbol{u}_{N-1}\}$. In the ideal chain case, as there are no constraints on the arrangements of the segments due to excluded volume effects, each bond vector is chosen randomly from the z vectors directed towards the nearest neighbor lattice sites. The orientation of adjacent bond vectors are therefore statistically independent.

This is expressed mathematically as follows:

$$\langle \boldsymbol{u}_i \cdot \boldsymbol{u}_j \rangle = \langle \hat{\boldsymbol{u}}_i \cdot \hat{\boldsymbol{u}}_j \rangle b_0^2 = \begin{cases} \langle |\hat{\boldsymbol{u}}_i|^2 \rangle b_0^2 = b_0^2 & (i = j)\ , \\ \langle \boldsymbol{u}_i \rangle \cdot \langle \boldsymbol{u}_j \rangle = \boldsymbol{0} \cdot \boldsymbol{0} = 0 & (i \neq j)\ , \end{cases} \tag{2.5}$$

where $\hat{\boldsymbol{u}}_i \equiv \boldsymbol{u}_i / |\boldsymbol{u}_i|$ is the unit vector in the direction parallel to $\boldsymbol{u}_i$, b_0 is the bond length (constant), and $\langle\ \rangle$ denotes the statistical average over many possible chain conformations in the equilibrium state, i.e., the ensemble

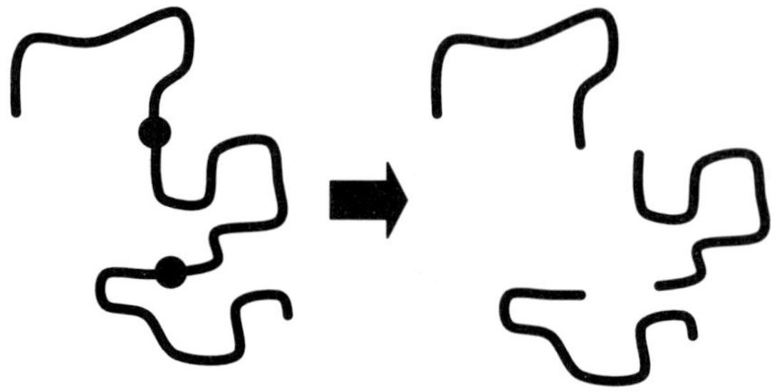

Fig. 2.9. Decomposing an ideal chain into a set of statistically independent subchains

average. In the derivation of (2.5), we used the relation $\langle \hat{\boldsymbol{u}}_i \rangle = \mathbf{0}$ due to the symmetry of the lattice structure.[4]

Equation (2.5) means that the directions of the bond vectors are statistically independent. Therefore, when an ideal chain is divided into two parts at a certain segment, the two resulting subchains are statistically independent. Repeating such divisions of the chain into subchains, an ideal chain can be decomposed into a set of statistically independent subchains, as shown in Fig. 2.9. This property of the ideal chain will be used in the self-consistent field theory introduced in Chap. 3.

Scaling of the End-to-End Distance

One of the measures of the spatial extension of a polymer chain is the distance between the two end segments of the chain. As shown in Fig. 2.10, the end-to-end vector is defined as

$$\boldsymbol{R} \equiv \boldsymbol{r}_N - \boldsymbol{r}_0 \,, \tag{2.6}$$

and the root mean square of the magnitude of this end-to-end vector

$$\sqrt{\langle |\boldsymbol{R}|^2 \rangle} = \sqrt{\langle |\boldsymbol{r}_N - \boldsymbol{r}_0|^2 \rangle} \tag{2.7}$$

is called the end-to-end distance. Here, note that the quantity $\langle \boldsymbol{R} \rangle$ is not suitable for the definition of the end-to-end distance because the average of any odd power of a vector quantity vanishes due to the isotropy of the system.

Using the property of the bond orientational correlations given by (2.5), the end-to-end distance defined by (2.7) is

[4] A bond vector $\boldsymbol{u}$ and its negative $-\boldsymbol{u}$ appear with the same probability due to the symmetry. In real polymers, there are often correlations between consecutive bond vectors. The treatment in such a realistic situation will be discussed in Sects. 2.5 and 2.6.

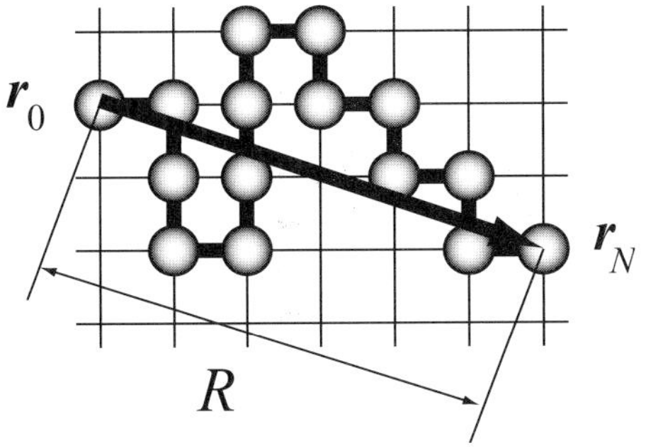

Fig. 2.10. End-to-end vector and end-to-end distance of a polymer chain

$$\begin{aligned}\langle|\boldsymbol{R}|^2\rangle &= \langle|\boldsymbol{r}_0-\boldsymbol{r}_N|^2\rangle \\ &= \langle|(\boldsymbol{r}_1-\boldsymbol{r}_0)+(\boldsymbol{r}_2-\boldsymbol{r}_1)+\cdots+(\boldsymbol{r}_N-\boldsymbol{r}_{N-1})|^2\rangle \\ &= \left\langle\left|\sum_{i=0}^{N-1}\boldsymbol{u}_i\right|^2\right\rangle \\ &= \sum_{i=0}^{N-1}\langle|\boldsymbol{u}_i|^2\rangle+\sum_i\sum_{j(j\neq i)}\langle\boldsymbol{u}_i\cdot\boldsymbol{u}_j\rangle = Nb_0^2\,. \end{aligned} \tag{2.8}$$

The average end-to-end distance of an ideal chain composed of N bonds is thus given by

$$\sqrt{\langle|\boldsymbol{R}|^2\rangle} = N^{1/2}b_0\,. \tag{2.9}$$

The important point in (2.9) is that the end-to-end distance is multiplied by a factor of $\sqrt{\lambda}$ when the chain length is multiplied by a factor of λ. Such a property is called a scaling property. The exponent 1/2 appearing in (2.9) is called the scaling exponent of an ideal chain. As will be discussed in the box entitled *Ideal Chains and Fractal Structure* in Sect. 2.3.1, the scaling property is related to the fractal nature of the chain conformation.

Another measure of the spatial extension of a polymer chain is the radius of gyration R_{G}. This quantity is defined as the root mean square of the distance between each segment and the center of mass of the chain (see Fig. 2.11). It is easy to show that the radius of gyration for $N \gg 1$ is given by (Exercise 2.1)

$$\langle R_{\mathrm{G}}^2\rangle \equiv \frac{1}{N+1}\left\langle\sum_{i=0}^{N}|\boldsymbol{r}_i-\boldsymbol{r}_{\mathrm{G}}|^2\right\rangle = \frac{1}{6}\frac{N(N+2)}{N+1}b_0^2 \sim \frac{1}{6}Nb_0^2\,. \tag{2.10}$$

Thus, as well as the end-to-end distance defined by (2.9), the radius of gyration depends on the chain length N as $N^{1/2}$. Figure 2.12 shows experimental data for the radius of gyration of polystyrene for various molecular

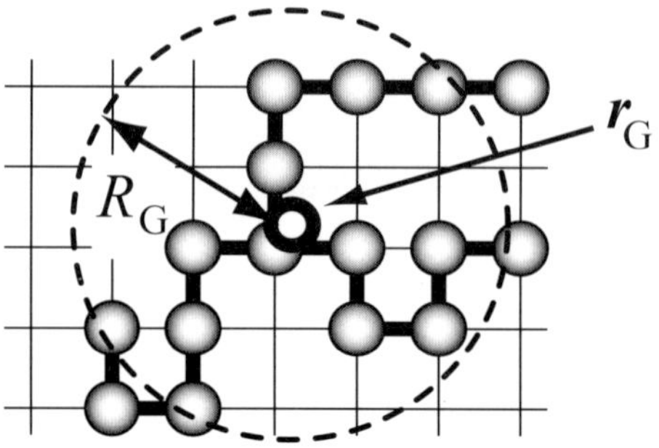

Fig. 2.11. In the case of branching polymers such as a comb polymer or a star polymer, the radius of gyration is a better measure of the spatial extension of the chain than the end-to-end distance. The *open circle* indicates the center of mass and the *large broken circle* corresponds to a sphere with radius R_G

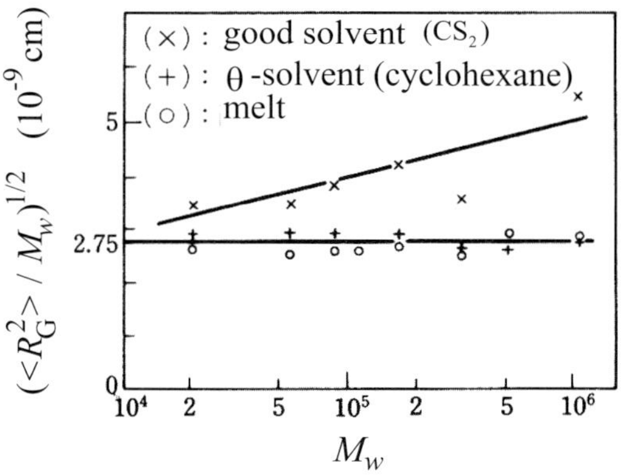

Fig. 2.12. Experimental results on the scaling behavior of the radius of gyration of polystyrene. The square of the gyration radius $\langle R_G^2 \rangle$ divided by the molecular weight M_w is plotted as a function of the molecular weight, which is proportional to the chain length. For the ideal chain case, this quantity is constant due to (2.10). Results for chains in a good solvent CS_2 (×), in a θ-solvent cyclohexane (+), and in a melt (○) are shown. Reprinted with permission from J.P. Cotton et al., Macromolecules **7**, 863 (1974). (© 1974, American Chemical Society)

weights. For polymers at the θ-temperature or for polymer melts,[5] the radius of gyration is proportional to the square root of the molecular weight. These experimental data thus demonstrate the scaling property of the ideal chains obtained above.

---

[5] A polymer melt is a liquid state of polymers without solvent. We discuss the properties of polymer melts in Sect. 2.3.4.

Probability Distribution of Chain Conformations

In terms of the probability distribution of the bond vectors, the statistical independence of the orientations of the bonds is expressed by

$$\begin{aligned} P_N(\{\boldsymbol{u}_i\}) &\equiv P_N(\boldsymbol{u}_0, \boldsymbol{u}_1, \ldots, \boldsymbol{u}_{N-1}) \\ &= P_1(\boldsymbol{u}_0)P_1(\boldsymbol{u}_1)\ldots P_1(\boldsymbol{u}_{N-1}) \,, \end{aligned} \tag{2.11}$$

where P_N denotes the joint probability distribution function for the N bonds and P_1 is the probability distribution function of a single bond. (see the box entitled *Basic Formulas of Probability Calculus* on p. 29). In deriving (2.11), we used the factorization property of the joint probability for statistically independent stochastic variables X and Y, i.e., $P_{\{X,Y\}}(x,y) = P_X(x)P_Y(y)$. We also assumed that all the bond vectors obey the same probability distribution function $P_1(\boldsymbol{u})$.

For the lattice model shown in Fig. 2.5a, the bond vector is chosen randomly from the z equivalent vectors directed from the central lattice site to the nearest neighbor lattice sites with equal probabilities. We can thus write $P_1(\boldsymbol{u})$ as follows (see Fig. 2.13a):

$$P_1(\boldsymbol{u}) = \begin{cases} 1/z & (\boldsymbol{u} \in \text{nearest neighbor vectors}) \,, \\ 0 & (\text{otherwise}) \,. \end{cases} \tag{2.12}$$

Basic Formulas of Probability Calculus

Let us consider two random variables X and Y. We denote the probability distribution function of X by $P_X(x)$ and the joint probability distribution function of X and Y by $P_{\{X,Y\}}(x,y)$. $P_X(x)$ is defined so that the probability of finding the value of X within the interval $[x, x + \mathrm{d}x]$ is equal to $P_X(x)\mathrm{d}x$. The joint probability $P_{\{X,Y\}}(x,y)$ is defined so that the probability of finding the values of X and Y within $[x, x+\mathrm{d}x]$ and $[y, y+\mathrm{d}y]$ is equal to $P_{\{X,Y\}}(x,y)\mathrm{d}x\mathrm{d}y$. These probability distribution functions have the following properties:

1. Any arguments of the joint probability distribution function can be eliminated by integrating the joint probability distribution function with respect to the argument:

$$\int P_{\{X,Y\}}(x,y)\mathrm{d}y = P_X(x) \,.$$

2. If X and Y are statistically independent, the joint probability distribution function can be written as a product of the distribution functions for X and Y, viz.,

$$P_{\{X,Y\}}(x,y) = P_X(x)P_Y(y) \,.$$

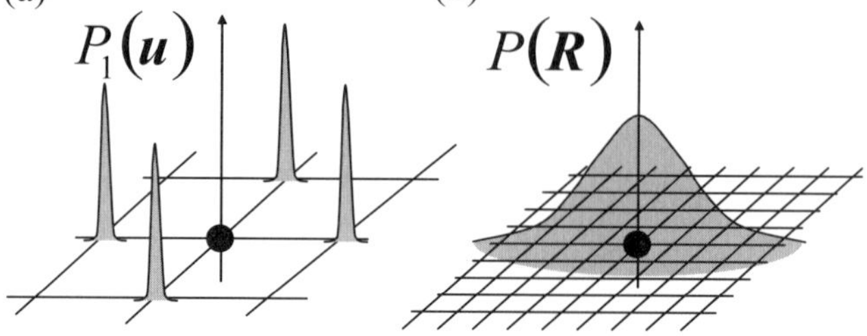

Fig. 2.13. (**a**) Probability distribution function for the bond vector and (**b**) the probability distribution function for the end-to-end vector for a lattice model of an ideal chain

Using the probability distribution function given by (2.11), the ensemble average $\langle\ \rangle$ appearing in (2.8) can be rewritten as

$$\langle * \rangle = \int d\boldsymbol{u}_0 \int d\boldsymbol{u}_1 \dots \int d\boldsymbol{u}_{N-1}\ *\ P_N(\{\boldsymbol{u}_i\}) , \tag{2.13}$$

where $*$ stands for an arbitrary physical quantity that depends on the bond vectors $\{\boldsymbol{u}_i\}$.

Probability Distribution for the End-to-End Vector

Using the ensemble average given by (2.13), let us calculate the probability of the event that the end-to-end vector equals a particular vector $\boldsymbol{R}$. First, we list the basic properties of the probability distribution functions.

Let $P_X(x)$ be the probability distribution function for a random variable X. Then the probability for an arbitrary function $f(X)$ to have the value a denoted $P_{f(X)}(a)$ is given by

$$P_{f(X)}(a) = \int_{-\infty}^{\infty} dx\, \delta\big(f(x) - a\big) P_X(x) , \tag{2.14}$$

where $\delta(x)$ is the Dirac delta-function. The above theorem means that an arbitrary function of X is also a random variable, and its probability distribution function is obtained by summing all the probabilities for X values that satisfy $f(X) = a$. The Dirac delta-function picks up the probability of such events among all possible events.

Basic Formulas for the Fourier Transformation

Let us consider a scalar function $f(\boldsymbol{r})$ of a 3-dimensional vector $\boldsymbol{r}$, where the domain of definition is assumed to be the whole 3-dimensional space. Then the Fourier transformation of $f(\boldsymbol{r})$ is defined by

$$\tilde{f}(\boldsymbol{q}) = \int \mathrm{d}\boldsymbol{r} f(\boldsymbol{r})\, \mathrm{e}^{-\mathrm{i}\boldsymbol{q}\cdot\boldsymbol{r}} . \tag{2.15}$$

Concerning this Fourier transformation, several useful formulas are listed here:

1. The inverse transformation of (2.15) is given by

$$f(\boldsymbol{r}) = \frac{1}{(2\pi)^3} \int \mathrm{d}\boldsymbol{q} \tilde{f}(\boldsymbol{q})\, \mathrm{e}^{\mathrm{i}\boldsymbol{q}\cdot\boldsymbol{r}} . \tag{2.16}$$

2. The Fourier transformation of a convolution integral of two arbitrary functions $f(\boldsymbol{r})$ and $g(\boldsymbol{r})$ is given by

$$\int \mathrm{d}\boldsymbol{r} \left[\int \mathrm{d}\boldsymbol{r}' f(\boldsymbol{r}-\boldsymbol{r}') g(\boldsymbol{r}') \right] \mathrm{e}^{-\mathrm{i}\boldsymbol{q}\cdot\boldsymbol{r}} = \tilde{f}(\boldsymbol{q}) \tilde{g}(\boldsymbol{q}) . \tag{2.17}$$

3. Concerning the convolution integral of three functions $f(\boldsymbol{r})$, $g(\boldsymbol{r})$ and $h(\boldsymbol{r})$ of $\boldsymbol{r}$, we obtain

$$\int \mathrm{d}\boldsymbol{r} \int \mathrm{d}\boldsymbol{r}' f(\boldsymbol{r}) g(\boldsymbol{r}') h(\boldsymbol{r}-\boldsymbol{r}') = \frac{1}{(2\pi)^3} \int \mathrm{d}\boldsymbol{q} \tilde{f}(\boldsymbol{q}) \tilde{g}(-\boldsymbol{q}) \tilde{h}(\boldsymbol{q}) . \tag{2.18}$$

Although it is not easy to treat the Dirac delta-function appearing in (2.14), it can be rewritten in a convenient form using its Fourier transform defined by

$$\int_{-\infty}^{\infty} \mathrm{d}q\, \mathrm{e}^{\mathrm{i}qx} = 2\pi\delta(x) . \tag{2.19}$$

Combining the general properties given in (2.14) and (2.19), the probability for the end-to-end vector to have the value $\boldsymbol{R}$ is given by

$$\begin{aligned} P(\boldsymbol{R}) &\equiv \int \mathrm{d}\boldsymbol{u}_0 \ldots \int \mathrm{d}\boldsymbol{u}_{N-1} \delta\left(\boldsymbol{R} - \sum_{i=0}^{N-1} \boldsymbol{u}_i \right) P_N(\{\boldsymbol{u}_i\}) \\ &= \frac{1}{(2\pi)^3} \int \mathrm{d}\boldsymbol{u}_0 \ldots \int \mathrm{d}\boldsymbol{u}_{N-1} \int \mathrm{d}\boldsymbol{q} \exp\left[\mathrm{i}\boldsymbol{q}\cdot \left(\boldsymbol{R} - \sum_{i=0}^{N-1} \boldsymbol{u}_i \right) \right] \prod_{j=0}^{N-1} P_1(\boldsymbol{u}_j) \\ &= \frac{1}{(2\pi)^3} \int \mathrm{d}\boldsymbol{q} \exp(\mathrm{i}\boldsymbol{q}\cdot\boldsymbol{R}) \prod_{j=0}^{N-1} \int \mathrm{d}\boldsymbol{u}_j \exp(-\mathrm{i}\boldsymbol{q}\cdot\boldsymbol{u}_j) P_1(\boldsymbol{u}_j) \\ &= \frac{1}{(2\pi)^3} \int \mathrm{d}\boldsymbol{q} \exp(\mathrm{i}\boldsymbol{q}\cdot\boldsymbol{R}) \left[\tilde{P}_1(\boldsymbol{q})\right]^N , \end{aligned} \tag{2.20}$$

where the last line is the inverse Fourier transform of $\left[\tilde{P}_1(\boldsymbol{q})\right]^N$. Here quantities with a tilde are Fourier-transformed quantities.

Central Limiting Theorem

Let us consider N independent random variables $\{X_1, X_2, \ldots, X_N\}$ that obey the same probability distribution function with mean value $\bar{X}$ and variance σ_X^2. Then the random variable Y defined by

$$Y = X_1 + X_2 + \cdots + X_N$$

obeys a probability distribution with mean value $\bar{Y} = N\bar{X}$ and variance $\sigma_Y^2 = N\sigma_X^2$. If $N \gg 1$, the probability distribution function for Y approaches the following Gaussian distribution:

$$P_Y(Y) = \frac{1}{\sqrt{2\pi\sigma_Y^2}} \exp\left[-\frac{1}{2\sigma_Y^2}\left(Y - \bar{Y}\right)^2\right].$$

This theorem is called the central limiting theorem. It indicates that, when N is large ($N \gg 1$), the width of the distribution of Y is proportional to $\sqrt{N}$, while its mean value is proportional to N. Thus the fluctuations in Y around its mean value decrease as $1/\sqrt{N}$, and eventually the fluctuations can be neglected in the very large limit of N ($N \to \infty$). Such a limit corresponds to the so-called thermodynamic limit.

Using (2.12), the probability distribution function for a single bond $P_1(\boldsymbol{u})$ of a lattice model is given by

$$P_1(\boldsymbol{u}) = \frac{1}{z}\sum_{\boldsymbol{a}} \delta(\boldsymbol{u} - \boldsymbol{a}), \tag{2.21}$$

where $\boldsymbol{a}$ runs over each of the z bond vectors pointing to the z nearest neighbor vectors, i.e., $\boldsymbol{a} = (\pm b_0, 0, 0)$, $(0, \pm b_0, 0)$, and $(0, 0, \pm b_0)$, for the three-dimensional cubic lattice. The Fourier transform of this distribution function $\tilde{P}_1(\boldsymbol{q})$ is given by

$$\tilde{P}_1(\boldsymbol{q}) = \int d\boldsymbol{u} P_1(\boldsymbol{u}) e^{-i\boldsymbol{q}\cdot\boldsymbol{u}} = \frac{1}{z}\sum_{\boldsymbol{a}} e^{-i\boldsymbol{q}\cdot\boldsymbol{a}}. \tag{2.22}$$

Gaussian Integrals

We present here several useful formulas concerning Gaussian integrals in 3-dimensional space:

1. Gaussian integral:

$$\int d\boldsymbol{r} \exp\left(-a|\boldsymbol{r} - \boldsymbol{b}|^2\right) = \left(\frac{\pi}{a}\right)^{3/2}, \tag{2.23}$$

where a is a positive constant and $\boldsymbol{b}$ is a constant vector.

2. Fourier transformation:

$$\int \mathrm{d}\boldsymbol{r} \exp\left(-a|\boldsymbol{r}-\boldsymbol{b}|^2\right)\mathrm{e}^{-\mathrm{i}\boldsymbol{q}\cdot\boldsymbol{r}} = \left(\frac{\pi}{a}\right)^{3/2} \exp\left(-\frac{|\boldsymbol{q}|^2}{4a} - \mathrm{i}\boldsymbol{q}\cdot\boldsymbol{b}\right) . \tag{2.24}$$

The universal properties of a polymer chain with $N \gg 1$ emerge on the length scale of the radius of gyration, which corresponds to the small wave number region. In such a case, we can expand (2.22) in a Taylor series in the wave number $\boldsymbol{q}$ and retain only the leading order terms. Using $z = 6$ for a 3-dimensional cubic lattice, we obtain

$$\begin{aligned}
\left[\tilde{P}_1(\boldsymbol{q})\right]^N &= \left\{\frac{1}{z}\sum_{\boldsymbol{a}}\left[1 - \mathrm{i}\boldsymbol{q}\cdot\boldsymbol{a} - \frac{1}{2}(\boldsymbol{q}\cdot\boldsymbol{a})^2 + \cdots\right]\right\}^N \\
&\sim \left[1 - \frac{b_0^2}{6}(q_x^2 + q_y^2 + q_z^2)\right]^N \\
&\sim \left(1 - \frac{b_0^2|\boldsymbol{q}|^2}{6}\right)^N \sim \exp\left(-\frac{Nb_0^2|\boldsymbol{q}|^2}{6}\right) ,
\end{aligned} \tag{2.25}$$

where $\boldsymbol{q} \equiv (q_x, q_y, q_z)$ and we have used the formula (2.23) for the Gaussian integral. Although we used the special value $z = 6$ for a cubic lattice in deriving (2.25), any other choice will lead to the same result. Substituting (2.25) into (2.20) and performing the inverse Fourier transform, we obtain the following Gaussian distribution for the end-to-end vector (see Fig. 2.13b):

$$P(\boldsymbol{R}) = \left(\frac{3}{2\pi N b_0^2}\right)^{3/2} \exp\left(-\frac{3|\boldsymbol{R}|^2}{2Nb_0^2}\right) . \tag{2.26}$$

Here one should note that the derivation of the Gaussian distribution given in (2.26) does not depend on any characteristic features of the lattice model. In deriving (2.25), the only approximation we have used is the long wavelength limit, where the wave vector $\boldsymbol{q}$ is small. In such a limit, any short range properties of the same order as the lattice constant b_0 can be neglected.

In fact, one can also derive the Gaussian distribution (2.26) starting from another model of the polymer chain in which the segments are connected by freely-rotating rigid rods (Exercise 2.2). This is a manifestation of the central limiting theorem which states that a sum of a large number of independent random variables tends to obey the Gaussian distribution. Thus the Gaussian nature of the end-to-end vector described by (2.26) is not a peculiar property of the model but is one of the universal properties of polymers.

2.2 Bead–Spring Model of Polymer Chain and Gaussian Chain Statistics

2.2.1 Coarse-Graining Procedure and Bead–Spring Model

Coarse-Graining Procedure

At the end of the last section, we showed that the probability distribution of the end-to-end vector obeys the Gaussian distribution for a very long ideal chain. As any two non-overlapping subchains of an ideal chain are statistically independent, the end-to-end vector of any subchain of an ideal chain obeys the Gaussian distribution as well. Let us consider a linear ideal chain composed of $N+1$ segments, numbered $0, 1, \ldots, N$. According to the above discussion, an arbitrary subchain of length m specified by the segments $i, i+1, \ldots, i+m$ exhibits the same ideal chain statistics. If m is large enough, the end-to-end vector of this subchain $\boldsymbol{R} = \boldsymbol{r}_{i+m} - \boldsymbol{r}_i$ obeys the following Gaussian distribution, similar to (2.26):

$$P(\boldsymbol{R}) = \left(\frac{3}{2\pi m b_0^2}\right)^{3/2} \exp\left(-\frac{3|\boldsymbol{R}|^2}{2mb_0^2}\right) . \tag{2.27}$$

To describe the physical properties on large scales, it is useful to construct the so-called coarse-grained model by eliminating the information concerning microscopic details. As shown in Fig. 2.14, we can define a coarse-grained segment by grouping several segments of the lattice model. We also define a coarse-grained chain as a sequence of these coarse-grained segments. If the coarse-grained segment is composed of a large number of the original segments, the distance between adjacent coarse-grained segments obeys the Gaussian distribution given by (2.27). This model can be a useful starting point for studying the physical properties of polymers.

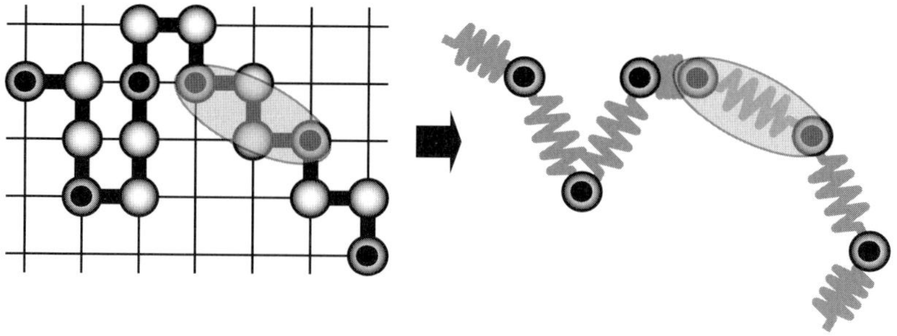

Fig. 2.14. Transformation from a lattice model to a bead–spring model by regarding several segments as a single coarse-grained segment. In this example, three segments of the lattice model form a coarse-grained segment. The positions of the centers of mass of every three original segments are indicated by *circles*, which are defined as the positions of the coarse-grained segments

Bead–Spring Model

In this section, we introduce a coarse-grained model for a polymer chain using the Gaussian property of the coarse-grained polymer chain discussed in the last section.

Relationship Between Lattice Model and Bead–Spring Model

At first glance, the Hamiltonians for the lattice model and the bead–spring model appear to be very different. In the bead–spring model, the bond that connects the segments is the harmonic spring given by (2.28). On the other hand, in the lattice model, there is no apparent spring potential between the segments. Instead, there is a constraint on the arrangement of the segments, i.e., a pair of segments that are connected by a chemical bond should be located on nearest neighbor lattice points. The chemical bonds are therefore introduced into the bead–spring model as a spring potential (enthalpic contribution), while they are introduced into the lattice model as an entropy associated with the constraint on the arrangement of the segments (entropic contribution). Because the final equilibrium state of the chain is determined by the free energy, composed of the enthalpic contribution E and the entropic contribution S according to

$$F = E - TS ,$$

we can understand that the two apparently different treatments of the chemical bonds will play essentially the same role in the canonical ensemble determined by the free energy F.

Firstly, we consider the size of the coarse-grained segment made of m original segments. The measure of the end-to-end distance of a subchain composed of m original segments is about $\sqrt{m}b_0$, as given by (2.27). This value corresponds to the size of the coarse-grained segment, which we denote by b $(= \sqrt{m}b_0)$. Secondly, we discuss the properties of the coarse-grained bonds connecting the coarse-grained segments. The probability distribution of the coarse-grained bond vector $\boldsymbol{R}$ is given by (2.27). Identifying this probability distribution with the canonical distribution of a harmonic spring, we can regard the coarse-grained bond as a spring with the harmonic potential

$$U_{\mathrm{bond}}(\boldsymbol{R}) = \frac{3k_{\mathrm{B}}T}{2b^2}|\boldsymbol{R}|^2 , \tag{2.28}$$

where b is the size of the coarse-grained segment obtained above. The polymer chain can thus be modeled by a sequence of coarse-grained segments connected by the harmonic spring (2.28) (see Fig. 2.15). This model is called the bead–spring model.

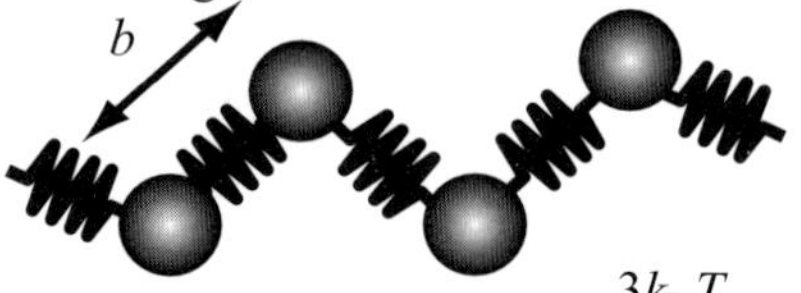

Fig. 2.15. Schematic picture of the bead–spring model

The remarkable point about the bond potential given in (2.28) is the fact that the spring constant depends on the temperature. This is because the spring of the bead–spring model originates from the change in the conformation entropy. Using such a spring potential, it is easy to show that the average of the square of the bond length is given by b^2 as follows:

$$\langle |\boldsymbol{u}|^2 \rangle = \frac{\displaystyle\int \mathrm{d}\boldsymbol{u} |\boldsymbol{u}|^2 \exp\left[-\beta U_{\mathrm{bond}}(\boldsymbol{u})\right]}{\displaystyle\int \mathrm{d}\boldsymbol{u} \exp\left[-\beta U_{\mathrm{bond}}(\boldsymbol{u})\right]} = b^2 \ . \tag{2.29}$$

Therefore, we can identify b as the average bond length of the coarse-grained model. This quantity b is often called the effective bond length.

Now let us define the Hamiltonian of the bead–spring model. We consider a linear chain composed of the same type of $N+1$ segments connected by the potential (2.28). We number each segment $i = 0, 1, \ldots, N$ from one end to the other. We assume that the solvent is so viscous that the kinetic energy can be neglected. Then the Hamiltonian of this polymer chain is given by

$$\mathcal{H} = \frac{3k_BT}{2b^2} \sum_{i=0}^{N-1} |\boldsymbol{r}_i - \boldsymbol{r}_{i+1}|^2 + \mathcal{W}(\boldsymbol{r}_0, \boldsymbol{r}_1, \ldots, \boldsymbol{r}_N) \equiv \mathcal{H}_0 + \mathcal{W} \ , \tag{2.30}$$

where $\boldsymbol{r}_i$ is the position vector of the ith segment. The term denoted by $\mathcal{W}$ is the potential energy, which depends on the chain conformation $\{\boldsymbol{r}_i\}$, including excluded volume interactions between segments or the effects of an external field such as an electric field.

Statistical Independence of Subchains

As shown in Fig. 2.9, an ideal chain can be decomposed into statistically independent subchains. Can the same decomposition be made for a chain with the interaction potential energy given by the Hamiltonian (2.30)?

In order for a pair of subchains to be statistically independent, the joint probability distribution function for their conformations, denoted by Γ_1 and

Γ_2, should be written as a product of the probability distributions for the conformations of the individual subchains (see the box entitled *Basic Formulas of Probability Calculus* on p. 29). This condition is satisfied if the Hamiltonian for the combined system of these two subchains $\mathcal{H}_{1+2}$ can be written as the sum of the Hamiltonians for the individual subchains, viz., $\mathcal{H}_{1+2} = \mathcal{H}_1 + \mathcal{H}_2$. This can easily be confirmed as follows:

$$P_{1+2}(\Gamma_1, \Gamma_2) = \text{const.} \times \mathrm{e}^{-\beta\mathcal{H}_{1+2}} = \text{const.} \times \mathrm{e}^{-\beta\mathcal{H}_1}\mathrm{e}^{-\beta\mathcal{H}_2} = P_1(\Gamma_1)P_2(\Gamma_2) \,.$$

If there is an interaction upon contact between segments, the Hamiltonian $\mathcal{H}_{1+2}$ contains the interaction term

$$\mathcal{W}(\boldsymbol{r}_0, \ldots, \boldsymbol{r}_N) = \frac{k_\mathrm{B}Tv}{2} \sum_{i \neq j} \delta(\boldsymbol{r}_i - \boldsymbol{r}_j) \,,$$

where the parameter v corresponds to the excluded volume parameter. This interaction term obviously violates the above-mentioned condition for statistical independence. Thus, in a chain with contact interaction between segments, the subchains cannot be treated as statistically independent. This causes a difficulty in evaluating the conformational probability distribution of the chain. In order to avoid such a difficulty, we often rely on the so-called mean field approximation, in which the position of one of the two interacting segments i and j is first averaged over its equilibrium probability distribution (see Sect. 2.3.3 for details).

The bead–spring model defined by (2.30) is a useful starting point for studying the statistical mechanical properties of polymer chains.

2.3 Statistical Mechanical Theory of Equilibrium Conformations of a Gaussian Chain

Given the model Hamiltonian (2.30) of a polymer chain, we can calculate static equilibrium properties such as the probability distribution of the chain conformation, equilibrium correlations between segments, and so on. We first study how to calculate the statistical properties of the conformation of an ideal chain whose Hamiltonian is given by (2.30) without the interaction term $\mathcal{W}$. We then analyze excluded volume chains with segment interactions using various approximations.

2.3.1 Ideal Chain Statistics

As already defined for the lattice model, an ideal chain has no interaction between segments. Such an ideal chain possesses only the effects of segment connectivity. The same situation can also be realized in the bead–spring model.

In this model, an ideal chain is described by the Hamiltonian $\mathcal{H}_0$ in (2.30) without the interaction potential $\mathcal{W}$. In this section, we study the statistical properties of the ideal chain using the bead–spring model.

Probability Distribution of Conformations of an Ideal Chain

The Hamiltonian of the bead–spring model of an ideal chain is given by the first term on the right-hand side of (2.30):

$$\mathcal{H}_0 = \frac{3k_{\mathrm{B}}T}{2b^2} \sum_{i=0}^{N-1} |\boldsymbol{r}_i - \boldsymbol{r}_{i+1}|^2 \, . \tag{2.31}$$

Let us use the canonical ensemble where the temperature, the total system volume, and the total number of particles in the system are fixed. The probability of finding the conformation $\{\boldsymbol{r}_0, \ldots, \boldsymbol{r}_N\}$ of an ideal chain is given using (2.31):

$$P(\{\boldsymbol{r}_i\}) = \frac{1}{Z_0} \exp\left(-\beta\mathcal{H}_0\right) = \frac{1}{Z_0} \exp\left(-\frac{3}{2b^2} \sum_{i=0}^{N-1} |\boldsymbol{r}_i - \boldsymbol{r}_{i+1}|^2\right) , \tag{2.32}$$

where Z_0 is the partition function of the chain defined by

$$Z_0 = \int \mathrm{d}\boldsymbol{r}_0 \ldots \int \mathrm{d}\boldsymbol{r}_N \exp\Big[-\beta\mathcal{H}_0(\boldsymbol{r}_0, \ldots, \boldsymbol{r}_N)\Big] \, . \tag{2.33}$$

The integral in (2.33) is easily performed using the variable transformation from the position vectors of the segments $\{\boldsymbol{r}_i\}$ to the bond vectors $\{\boldsymbol{u}_i\}$ and the position of the end segment $\boldsymbol{r}_0$. Using the definition of (2.1), we can check that the Jacobian of this transformation is unity, so that

$$\begin{aligned} \mathrm{d}\boldsymbol{r}_0 \ldots \mathrm{d}\boldsymbol{r}_N &= \left| \frac{\partial(\boldsymbol{r}_0, \ldots, \boldsymbol{r}_N)}{\partial(\boldsymbol{r}_0, \boldsymbol{u}_0, \ldots, \boldsymbol{u}_{N-1})} \right| \mathrm{d}\boldsymbol{r}_0 \mathrm{d}\boldsymbol{u}_0 \ldots \mathrm{d}\boldsymbol{u}_{N-1} \\ &= \mathrm{d}\boldsymbol{r}_0 \mathrm{d}\boldsymbol{u}_0 \ldots \mathrm{d}\boldsymbol{u}_{N-1} \, . \end{aligned} \tag{2.34}$$

Applying this relation and the formula (2.23) for the Gaussian integral, we can perform the integral in (2.33) to obtain

$$Z_0 = \mathcal{V} \left(\frac{2\pi b^2}{3}\right)^{3N/2} , \tag{2.35}$$

where $\mathcal{V}$ is the volume of the system over which the center of mass of the chain can wander. This quantity Z_0 corresponds to the total number of possible conformations realized by an ideal chain composed of $N+1$ segments, multiplied by the extra factor $\mathcal{V}$ which accounts for the free volume for the center of mass of the chain.

Important Formulas for Gaussian Statistics

Consider a one-dimensional Gaussian random variable X with mean value $\langle X \rangle = \bar{X}$ and variance $\langle (X - \bar{X})^2 \rangle = \sigma_X^2$. The probability distribution function of X is given by

$$P_X(X) = \frac{1}{\sqrt{2\pi\sigma_X^2}} \exp\left[-\frac{1}{2\sigma_X^2}(X - \bar{X})^2\right] .$$

We can assume that the mean value is zero without loss of generality by redefining $X - \bar{X}$ as X. Then the following formulas can be derived:

1. Wick's Theorem:

$$\langle X_1 X_2 \ldots X_{2n} \rangle = \sum_{\substack{\text{all possible combinations} \\ \text{of pairs}}} \langle X_{i_1} X_{i_2} \rangle \langle X_{i_3} X_{i_4} \rangle \ldots \langle X_{i_{2n-1}} X_{i_{2n}} \rangle ,$$

where the set of indices $\{i_1, \ldots, i_{2n}\}$ is a permutation of $\{1, \cdots, 2n\}$ and the sum is taken over all possible permutations.

2. Statistical average of the Fourier transform of X :

$$\langle \exp(-\mathrm{i}qX) \rangle = \exp\left(-\frac{1}{2}q^2 \langle X^2 \rangle\right) = \exp\left(-\frac{1}{2}\sigma_X^2 q^2\right) .$$

By applying the formula (2.14) to the joint probability distribution function of all the segments, we can derive the probability distribution functions of arbitrary subchains or the probability distribution function under given constraints where certain degrees of freedom are fixed. For example, the probability for the end-to-end vector $\boldsymbol{r}_N - \boldsymbol{r}_0$ to have a specified value $\boldsymbol{R}$ is obtained using (2.19) and (2.32) as

$$\begin{aligned} P(\boldsymbol{R}) &= \int \mathrm{d}\boldsymbol{r}_0 \ldots \int \mathrm{d}\boldsymbol{r}_N \delta(\boldsymbol{r}_N - \boldsymbol{r}_0 - \boldsymbol{R}) P(\{\boldsymbol{r}_i\}) \\ &= \left(\frac{3}{2\pi N b^2}\right)^{3/2} \exp\left(-\frac{3}{2Nb^2}|\boldsymbol{R}|^2\right) . \end{aligned} \tag{2.36}$$

The probability distribution function for the end-to-end vector of a chain with arbitrary chain length N is thus a Gaussian distribution.

The free energy $F_0(\boldsymbol{R})$ of an ideal chain whose end-to-end vector is fixed at $\boldsymbol{R}$ [or equivalently the conformational entropy $S_0(\boldsymbol{R})$] is obtained using (2.36):

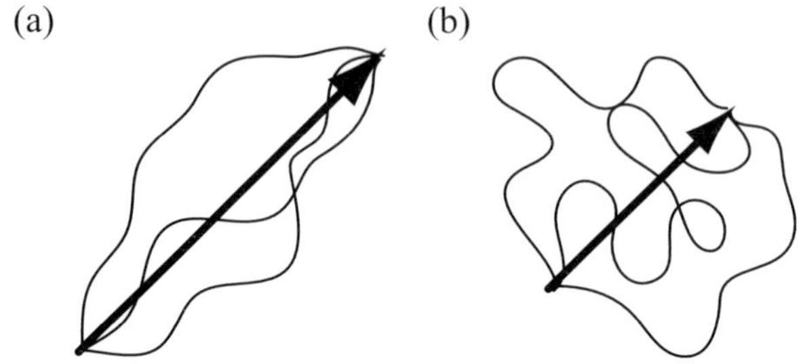

Fig. 2.16. Relationship between the number of possible conformations and the conformational entropy of an ideal chain. (**a**) A chain with a large end-to-end distance, and (**b**) a chain with small end-to-end distance. In the case of the stretched chain shown in (**a**), the number of possible conformations becomes fewer. The conformational entropy is thus small, which leads to a larger free energy

$$\begin{aligned} F_0(\boldsymbol{R}) &= -TS_0(\boldsymbol{R}) \\ &= -k_{\mathrm{B}}T \ln \begin{pmatrix} \text{Number of possible conformations} \\ \text{given the end-to-end distance} \end{pmatrix} \\ &= -k_{\mathrm{B}}T \ln \left[\begin{pmatrix} \text{Number of possible conformations} \\ \text{of an ideal chain without constraints} \end{pmatrix} \times P(\boldsymbol{R}) \right] \\ &= -k_{\mathrm{B}}T \ln \left[\frac{Z_0}{\mathcal{V}} \times P(\boldsymbol{R}) \right] = \frac{3k_{\mathrm{B}}T}{2Nb^2} |\boldsymbol{R}|^2 + C(N) \,, \end{aligned} \tag{2.37}$$

where $C(N)$ is a constant that depends only on the chain length N. In deriving (2.37), we have used the fact that the total number of possible conformations of an unconstrained ideal chain is given by (2.35), except for the factor $\mathcal{V}$ coming from the translational degrees of freedom of the center of mass. This result shows that the free energy of an ideal chain is proportional to the square of the end-to-end distance $|\boldsymbol{R}|$. Thus the separation between the two end points of an ideal chain of length N can be identified with the ends of a harmonic spring with spring constant $3k_{\mathrm{B}}T/(Nb^2)$ (see Fig. 2.16).

Ideal Chains and Fractal Structures

As discussed in connection with (2.26), the probability distribution of the end-to-end vector of a lattice model of an ideal chain becomes approximately Gaussian only in the long-chain limit. The bead–spring model of an ideal chain discussed here, however, shows the Gaussian nature from the scale of a single bond [see (2.28)] to the scale of the entire chain [see (2.36)]. Such a feature is said to be scale free and referred to as a scaling property. Chains that show the property given by (2.36) are called Gaussian chains.

Ideal Chains and Fractal Structures

A scale-free structure looks similar on any length scale, i.e., if we enlarge a small portion of a scale-free structure, it appears to be the same as the original one. Such a pattern is said to be self-similar and the structure is called a fractal structure. Here we show that a spatial or temporal self-similar (fractal) random structure has a power law distribution.

Let us denote the probability distribution function of a self-similar random variable x by $f(x)$. Due to the self-similar nature, the functional form of $f(x)$ should be left unchanged within a constant factor when the variable x is multiplied by an arbitrary expansion factor λ. Thus, for any value of λ, the following relation should hold:

$$f(\lambda x) = \alpha(\lambda) f(x) ,$$

where the function $f(x)$ only changes by a constant factor $\alpha(\lambda)$ which depends on λ. Differentiating the above equation with respect to λ and substituting $\lambda = 1$, we obtain the ordinary differential equation

$$x f'(x) = \alpha'(1) f(x) ,$$

where the prime denotes differentiation with respect to the argument. Solving this ordinary differential equation, we immediately obtain the power law function

$$f(x) = \text{const.} \times x^{\alpha'(1)} .$$

Figure 2.18 shows two typical examples of fractal curves, i.e., the Koch curve and the trace of a Brownian particle.

The fractal dimension of a spatially distributed physical quantity is defined using its real space correlation function $C(r)$, which is proportional to the probability of finding the quantity at two distant points separated by the distance r. The above argument requires this correlation function to be a power-law function of the distance r. Then the fractal dimension D is defined as $C(r) \sim r^{D-d}$, where d is the spatial dimension.

The scaling property of the Gaussian chain means that the statistical properties of the conformation do not change from smaller scales to larger scales. Such a structure is called a fractal structure (see the box entitled *Ideal Chains and Fractal Structures*).

We should note that the total length of a scale-free chain depends on the size of the unit used to measure the length. Because a scale-free curve has a zigzag structure on all length scales, a shorter unit can detect smaller structures. Thus, the shorter the unit becomes, the longer the measured total

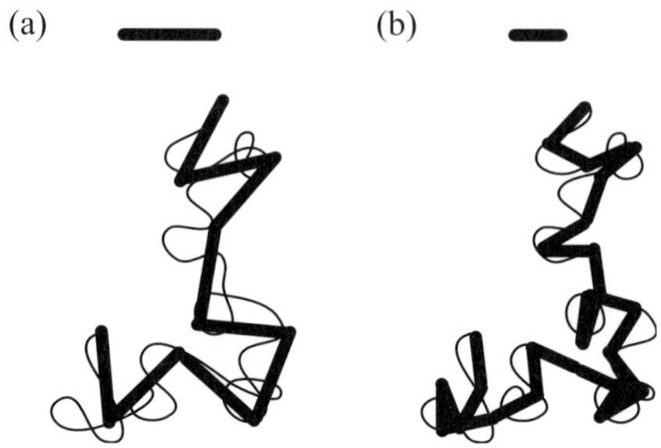

Fig. 2.17. Measurements of the length of a scale-free chain using two different units (the *bars* at the top of each figure). (**a**) Using a longer unit, small-scale structures cannot be detected and the measured total length of the chain becomes shorter. (**b**) Using a shorter unit, the small-scale structures can be well detected and the measured total length becomes longer

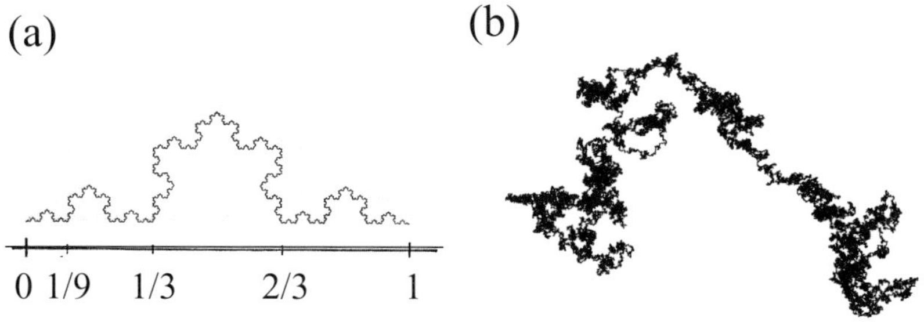

Fig. 2.18. Two examples of fractal curves. (**a**) The Koch curve and (**b**) the trace of a Brownian particle (a random walk). The Koch curve has a regular fractal structure. Any portion of the curve has exactly the same structure as the original curve. On the other hand, the statistical properties of the random walk are preserved when part of it is enlarged. (**a**) Reprinted with permission from *Fractal* by H. Takayasu (© 1986, Asakura Shoten) in Japanese. (**b**) Courtesy of H. Tasaki

chain length is. Therefore, the total chain length of a Gaussian chain is not well-defined until we define the unit length[6] (see Fig. 2.17).

[6] A real polymer chain has an upper limit for its extension. In contrast, a Gaussian chain can be stretched up to any length by imposing a sufficiently large extensional force.

2.3.2 Correlation Functions and Scattering Functions of an Ideal Chain

Scattering Experiments

Scattering experiments such as neutron scattering, X-ray scattering, and light scattering experiments are effective methods for observing the conformation of polymer chains experimentally. There are two types of scattering experiment: one involves elastic scattering, in which there is no energy exchange between the probe (neutrons or photons) and the specimen, and the other involves inelastic scattering, in which the probe and the specimen exchange energy. In order to study static structures such as the probability distributions of chain conformations, elastic scattering experiments are suitable.

In an elastic scattering experiment, a neutron or X-ray beam is applied to the specimen. The intensity of the scattered neutrons or scattered photons is detected at various angles. This directly reflects the correlations of the fluctuations in the distributions of the atomic nuclei or the electrons in the specimen. Schematic illustrations of scattering experiments are shown in Fig. 2.19.

Using elastic scattering experiments, the Fourier components of the correlation function of the density fluctuations at two different points in the specimen can be measured. As an example, we show in Fig. 2.20 the result of an elastic neutron scattering experiment on a single isolated chain in a polymer melt. In order to observe the conformation of a single chain in a melt,

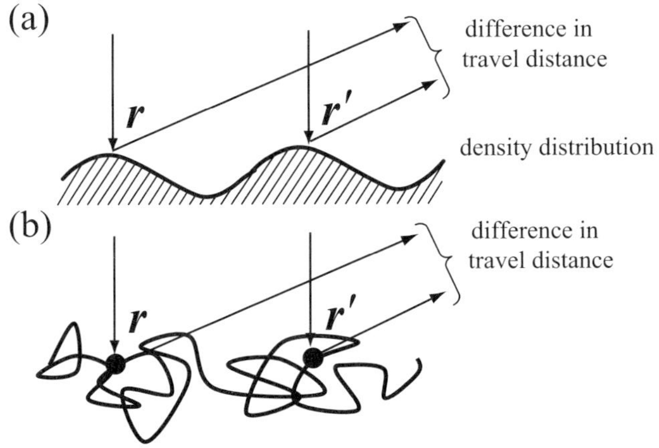

Fig. 2.19. Schematic illustrations of elastic neutron or X-ray scattering experiments. (**a**) Scattering from continuously distributed density fluctuations, and (**b**) scattering from the segments of a polymer. When the difference in the distances traveled by the probes is an integer multiple of the wavelength λ, the scattered beam has a large intensity

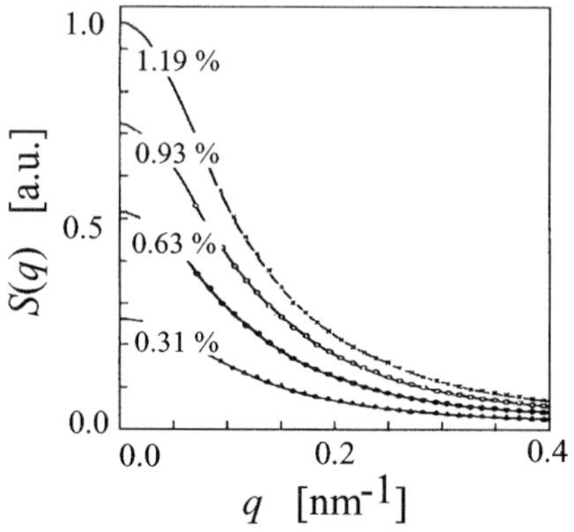

Fig. 2.20. Scattering intensity of labeled polymethylmethacrylate (PMMA), dilutely dispersed in a melt of unlabeled PMMA, obtained by an elastic neutron scattering experiment. The *continuous curve* is the theoretical prediction using the ideal chain statistics given by (2.42), i.e., the so-called Debye function, which is modified by taking into account the polydispersity effect (the distribution of the molecular weight). *Numbers* show the density of the labeled PMMA. Reprinted with permission from R. Kirste et al., Polymer **16**, 120 (1975). (© 1975, Elsevier Science)

one adds a very small amount of labeled polymers, in which the protons are replaced by deuterons, to a melt of unlabeled polymers. As the scattering length of a neutron by a proton is different from that by a deuteron, the scattering of the labeled polymer chains can be distinguished from those of the unlabeled chains. The agreement between the experimental data and the theoretical predictions for the ideal chain (described later) shown in Fig. 2.20 implies that ideal chain statistics can be applied to the melt state. This is due to screening effects that will be discussed in Sect. 2.3.4.

Correlation Functions and Scattering Functions

We now calculate the two-point correlation function of an ideal chain. The two-point correlation function of the segment density is defined by

$$\Phi_0(\boldsymbol{r}, \boldsymbol{r}') \equiv \frac{1}{N} \sum_i \sum_j \langle \delta(\boldsymbol{r} - \boldsymbol{r}_i) \delta(\boldsymbol{r}' - \boldsymbol{r}_j) \rangle . \tag{2.38}$$

If there is no external inhomogeneous field imposed on this chain, the center of mass of the chain can be located equally anywhere in the system. For this reason, the two-point correlation function $\Phi_0(\boldsymbol{r}, \boldsymbol{r}')$ should have the translational invariance described by

$$\Phi_0(\boldsymbol{r}, \boldsymbol{r}') = \frac{1}{\mathcal{V}} \Phi_0(\boldsymbol{r} - \boldsymbol{r}') , \tag{2.39}$$

where $\mathcal{V}$ is the system volume which arises from the translational degree of freedom of the center of mass of the chain which can wander the region with volume $\mathcal{V}$. This property will actually be confirmed below.

Using the probability distribution for the conformation of an ideal chain given by (2.32), we obtain

$$\begin{aligned}\Phi_0(\boldsymbol{r},\boldsymbol{r}') &= \frac{1}{N}\sum_i\sum_j \int \mathrm{d}\boldsymbol{r}_0 \ldots \int \mathrm{d}\boldsymbol{r}_N \delta(\boldsymbol{r}-\boldsymbol{r}_i)\delta(\boldsymbol{r}'-\boldsymbol{r}_j)P(\{\boldsymbol{r}_i\}) \\ &= \frac{1}{N}\sum_i\sum_j \frac{1}{Z_0(|j-i|)}\int \mathrm{d}\boldsymbol{r}_i \int \mathrm{d}\boldsymbol{r}_{i+1}\ldots\int \mathrm{d}\boldsymbol{r}_j \\ &\quad \times\delta(\boldsymbol{r}-\boldsymbol{r}_i)\delta(\boldsymbol{r}'-\boldsymbol{r}_j)\exp\left(-\frac{3}{2b^2}\sum_{k=i}^{j-1}|\boldsymbol{r}_k-\boldsymbol{r}_{k+1}|^2\right) , \end{aligned} \tag{2.40}$$

where $Z_0(m)$ is the partition function of a subchain with length m.

In deriving the second equality, we performed the integrations with respect to the coordinates except for $\{\boldsymbol{r}_i,\ldots,\boldsymbol{r}_j\}$ because non-overlapping subchains of an ideal chain are statistically independent. The integral in the last line is the conformational probability distribution of an ideal chain whose ends are fixed at $\boldsymbol{r}_i$ and $\boldsymbol{r}_j$. This is given by the Gaussian distribution in (2.36). Substituting (2.35) and (2.36) into (2.40) and taking the continuum limit, where the discrete segment index i is replaced by a continuous variable s, we obtain

$$\begin{aligned}\Phi_0(\boldsymbol{r},\boldsymbol{r}') &= \frac{1}{N\mathcal{V}}\sum_i\sum_j\left(\frac{3}{2\pi|j-i|b^2}\right)^{3/2}\exp\left(-\frac{3|\boldsymbol{r}-\boldsymbol{r}'|^2}{2|j-i|b^2}\right) \\ &\sim \frac{1}{N\mathcal{V}}\int_0^N \mathrm{d}s\int_0^N \mathrm{d}s'\left(\frac{3}{2\pi|s-s'|b^2}\right)^{3/2}\exp\left(-\frac{3|\boldsymbol{r}-\boldsymbol{r}'|^2}{2|s-s'|b^2}\right) \\ &\equiv \frac{1}{\mathcal{V}}\Phi_0(\boldsymbol{r}-\boldsymbol{r}') ,\end{aligned} \tag{2.41}$$

where we used the fact that Φ_0 is a function of $\boldsymbol{r}-\boldsymbol{r}'$ because the system has translational invariance with respect to the center of mass of the chain. Fourier transforming both sides of (2.41) with respect to the coordinate $\boldsymbol{r}-\boldsymbol{r}'$, we obtain

$$\begin{aligned}\tilde{\Phi}_0(\boldsymbol{q}) &= \frac{1}{N}\int_0^N \mathrm{d}s\int_0^N \mathrm{d}s' \exp\left(-\frac{1}{6}b^2|s-s'||\boldsymbol{q}|^2\right) \\ &= \frac{2N}{x^2}\left(\mathrm{e}^{-x}-1+x\right) ,\end{aligned} \tag{2.42}$$

where we have defined $x \equiv (1/6)Nb^2|\boldsymbol{q}|^2 = R_G^2|\boldsymbol{q}|^2$ and R_G is the radius of gyration defined by (2.10). As shown below, this $\tilde{\Phi}_0(\boldsymbol{q})$ corresponds to the scattering intensity measured in the scattering experiment. Equation (2.42) is thus the scattering intensity from an isolated ideal chain. Such a scattering intensity from an ideal chain is called the Debye function.

Physical Meaning of Two-Point Correlation Functions

The two-point correlation function defined by (2.38) is proportional to the probability of finding segments at two points $\boldsymbol{r}$ and $\boldsymbol{r}'$ simultaneously. If the probability distributions of the segment density at $\boldsymbol{r}$ and $\boldsymbol{r}'$ ($\boldsymbol{r} \neq \boldsymbol{r}'$) are statistically independent, the ensemble average in (2.38) can be decomposed into a product of the averages at $\boldsymbol{r}$ and $\boldsymbol{r}'$, and we obtain

$$\Phi_0(\boldsymbol{r}, \boldsymbol{r}') \approx \frac{1}{N} \sum_{i>j} \langle \delta(\boldsymbol{r} - \boldsymbol{r}_i) \rangle \langle \delta(\boldsymbol{r}' - \boldsymbol{r}_j) \rangle = \frac{N}{2} \phi(\boldsymbol{r}) \phi(\boldsymbol{r}') ,$$

where $\phi(\boldsymbol{r})$ is the segment density at $\boldsymbol{r}$. The important part of the two-point correlation function is the contribution that cannot be expressed in terms of the product of one-point distribution functions $\phi(\boldsymbol{r})$. Such a contribution corresponds to the amount of excess density distribution at point $\boldsymbol{r}$ induced by the segment fluctuation at another point $\boldsymbol{r}'$. Even for an ideal chain system, if we select two points $\boldsymbol{r}$ and $\boldsymbol{r}'$ within the radius of gyration of the chain, there must be a certain probability of having chain conformations that go through these two points simultaneously, which contribute to the correlations between the two points.

It can be shown that the scattering intensity measured in a scattering experiment is proportional to the Fourier transformation of the two-point correlation function $\Phi_0(\boldsymbol{r}, \boldsymbol{r}')$ defined by (2.38).

As shown in Fig. 2.19, consider a plane wave of the probe (X-ray, neutron beam, etc.) scattered by the segment density fluctuation of the specimen. The segment density fluctuation is described microscopically by

$$\delta\hat{\phi}(\boldsymbol{r}; \Gamma) = \sum_i \delta(\boldsymbol{r} - \boldsymbol{r}_i) - \phi_0 , \tag{2.43}$$

where $\Gamma \equiv \{\boldsymbol{r}_0, \ldots, \boldsymbol{r}_N\}$ is the microscopic phase space of the chain and ϕ_0 is the average segment density. The hat symbol indicates that this quantity is a microscopic quantity that depends on the phase space Γ.

The wave function of the scattered beam at a certain steric angle consists of a superposition of scattered waves from different points. This is expressed by the convolution of the segment density fluctuation and the plane waves as

$$\int \mathrm{d}\boldsymbol{r} \delta\hat{\phi}(\boldsymbol{r}; \Gamma) \mathrm{e}^{-\mathrm{i}\boldsymbol{q}\cdot\boldsymbol{r}} , \tag{2.44}$$

where $\boldsymbol{q}$ is the difference in the wave vectors between the incident beam and the scattered beam. This expression means that the scattered wave function is the Fourier transformation of the density fluctuation $\delta\hat{\phi}(\boldsymbol{r}; \Gamma)$.

The scattering intensity is the probability of detecting the scattered beam in the direction specified by $\boldsymbol{q}$. As this probability is given by the square of the wave function of the scattered probe, it is

$$S(\boldsymbol{q}) \equiv \left\langle |\delta\tilde{\phi}(\boldsymbol{q};\Gamma)|^2 \right\rangle . \tag{2.45}$$

This $S(\boldsymbol{q})$ is called the scattering function.

Substituting in the definition of the Fourier transformation and using the translational symmetry, it is easy to show that (2.45) can be expressed in terms of the Fourier transformation of the two-point correlation function (2.38) as follows:

$$\begin{aligned} S(\boldsymbol{q}) &= \left\langle |\delta\tilde{\phi}(\boldsymbol{q};\Gamma)|^2 \right\rangle \\ &= \int \mathrm{d}\boldsymbol{r} \int \mathrm{d}\boldsymbol{r}' \langle \delta\hat{\phi}(\boldsymbol{r};\Gamma)\delta\hat{\phi}(\boldsymbol{r}';\Gamma)\rangle \mathrm{e}^{-\mathrm{i}\boldsymbol{q}\cdot(\boldsymbol{r}-\boldsymbol{r}')} . \end{aligned} \tag{2.46}$$

Using the definitions (2.38), (2.39) and (2.43), we obtain

$$\begin{aligned} \frac{1}{\mathcal{V}} \Phi_0(\boldsymbol{r}-\boldsymbol{r}') &= \Phi_0(\boldsymbol{r},\boldsymbol{r}') \\ &= \frac{1}{N} \left\langle \delta\hat{\phi}(\boldsymbol{r};\Gamma)\delta\hat{\phi}(\boldsymbol{r}';\Gamma) \right\rangle + \frac{1}{N}\phi_0^2 . \end{aligned} \tag{2.47}$$

Combining (2.46) and (2.47), we finally obtain

$$\begin{aligned} S(\boldsymbol{q}) &= N \int \mathrm{d}\boldsymbol{r} \Phi_0(\boldsymbol{r}) \mathrm{e}^{-\mathrm{i}\boldsymbol{q}\cdot\boldsymbol{r}} - (2\pi)^3 \mathcal{V} \phi_0^2 \delta(\boldsymbol{q}) \\ &= N\tilde{\Phi}_0(\boldsymbol{q}) - (2\pi)^3 \mathcal{V}\phi_0^2 \delta(\boldsymbol{q}) , \end{aligned} \tag{2.48}$$

which relates the real-space two-point correlation function and the scattering function. This relation is called the Wiener–Khinchin relation.

Asymptotic Behavior of the Scattering Function

In (2.42), the two limiting cases $x \ll 1$ and $x \gg 1$ correspond to the short-range region and long-range regions compared with the radius of gyration of the chain R_G, i.e., $2\pi/|\boldsymbol{q}| \gg R_\mathrm{G}$ and $2\pi/|\boldsymbol{q}| \ll R_\mathrm{G}$, respectively. The asymptotic functional forms in these two limiting cases can be obtained from (2.42) as

$$\tilde{\Phi}_0(\boldsymbol{q}) = \begin{cases} N \sim \text{const.} & (x \ll 1 \ \text{ or } \ 2\pi/|\boldsymbol{q}| \gg R_\mathrm{G}) , \\ \dfrac{2N}{x} \sim |\boldsymbol{q}|^{-2} & (x \gg 1 \ \text{ or } \ 2\pi/|\boldsymbol{q}| \ll R_\mathrm{G}) . \end{cases} \tag{2.49}$$

In the case $2\pi/|\boldsymbol{q}| \gg R_\mathrm{G}$, the characteristic length of phenomena observed by the scattering experiment is much larger than the radius of gyration. In such a case, the internal structure of the coil of the chain cannot be resolved.

Therefore, the chain can be regarded as a delta function located at the center of mass of the chain, and this leads to the delta-function-like behavior of the correlation function. As the Fourier transform of such a delta-function gives a constant, the low-q region of the scattering function $\tilde{\Phi}_0(\boldsymbol{q})$ becomes constant.

On the other hand, in the case of $2\pi/|\boldsymbol{q}| \ll R_{\mathrm{G}}$, the scattering experiment detects only small-scale phenomena compared with the radius of gyration of the chain. As the local conformation of the Gaussian chain is a fractal structure with the same fractal dimension 2 as the random walk, one can observe the power law behavior described in the box entitled *Ideal Chains and Fractal Structures* on p. 41. The real space correlation function of a fractal structure with fractal dimension D ($0 \leq d - D \leq 1$) in d-dimensional space and its Fourier transform are given by

$$\Phi_0(\boldsymbol{r}) \sim |\boldsymbol{r}|^{D-d} \tag{2.50}$$

and

$$\Phi_0(\boldsymbol{q}) = \int \mathrm{d}\boldsymbol{r} \Phi_0(\boldsymbol{r})\, \mathrm{e}^{\mathrm{i}\boldsymbol{q}\cdot\boldsymbol{r}} \sim \int_0^\infty \mathrm{d}r\, r^{d-1} r^{D-d} \mathrm{e}^{\mathrm{i}\boldsymbol{q}\cdot\boldsymbol{r}} \sim |\boldsymbol{q}|^{-D}\,, \tag{2.51}$$

where $r = |\boldsymbol{r}|$ and the factor r^{d-1} in the second equation is the Jacobian relating the Cartesian coordinate to the polar coordinate. Thus the scattering function of the local structure of an ideal chain is obtained by substituting $d = 3$ and $D = 2$ into (2.50), which results in the second equation in (2.49).

As the scattering function discussed in this section is for the ideal chain without segment interactions, one may think that the results presented here cannot be applied to real systems. It is known, however, that the chains in concentrated polymer systems obey ideal chain statistics due to the screening effects that will be discussed in Sect. 2.3.4. In fact, the theoretically obtained scattering function (2.42) reproduces the experimentally observed scattering function of a single chain in a concentrated polymer system quite well (see Fig. 2.20). The scattering function (2.42) for an ideal chain is also useful for calculating the scattering function of realistic systems with segment interactions. As will be discussed in Sect. 4.1.3, the scattering function of a system of interacting chains can be calculated from the scattering function of the ideal chains given in (2.42) using the random phase approximation. In Chap. 4, this method is adopted to derive the Ginzburg–Landau model, which can be a useful tool for explaining qualitatively the behavior of concentrated polymer systems.

2.3.3 Statistical Mechanics of Chains with Interactions and Approximate Theories

Excluded Volume Interaction

In this section, we discuss the statistical properties of chains with interactions between the segments. A typical example of such interactions is the excluded volume interaction, which prevents two segments from occupying the same position. For this interaction, the potential term $\mathcal{W}$ in (2.30) is written as

$$\mathcal{W}(\boldsymbol{r}_0, \ldots, \boldsymbol{r}_N) = \frac{k_B T v}{2} \sum_i \sum_{j\ (j \neq i)} \delta(\boldsymbol{r}_i - \boldsymbol{r}_j) , \qquad (2.52)$$

where v is the excluded volume parameter introduced in Sect. 2.1.1. The excluded volume parameter v is a measure of the volume occupied by a segment.[7] The summation $\sum_i \sum_{i \neq j}$ should be taken over all the segment pairs i and j. Similarly to the lattice model case, the bead–spring model with the excluded volume interaction is called an excluded volume chain model.

As was discussed briefly in Sect. 2.1.1, the excluded volume parameter v accounts for the effective interaction between segments obtained by averaging out the degrees of freedom of the solvent molecules (see Fig. 2.21).

When we have the interaction term (2.52) in the Hamiltonian, the probability distribution of the chain conformation can no longer be calculated exactly and we have to rely on approximate theories. Typical approximate theories are the mean field theory and perturbation theory. In Chaps. 3 and 4, the mean field theory is combined with the self-consistent field theory or with the Ginzburg–Landau theory. The mean field theory is used to predict the behavior of concentrated many-chain systems quantitatively using a numerical technique. On the other hand, perturbation theory can give analytical results, although its applicability is in general very limited.

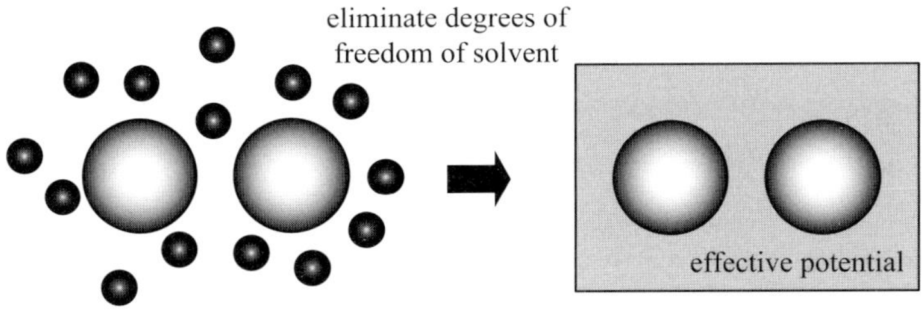

Fig. 2.21. The excluded volume parameter is a measure of the effective interaction between a pair of segments obtained by averaging over all possible arrangements of the solvent molecules under the constraint that the separation distance between the two segments is kept constant (the so-called adiabatic approximation)

[7] As the delta-function gives unity when it is integrated over the volume, it has the dimensions of inverse volume. Then, in order for the right-hand side of (2.52) to have the dimensions of energy, the excluded volume parameter v should have the dimensions of volume.

Mean Field Theory

Due to the nonlinear interactions in (2.52), the ensemble averages of products of segment positions such as $\langle \boldsymbol{r}_i \boldsymbol{r}_j \rangle$ $(i \neq j)$ cannot be evaluated exactly. We must then approximate them by neglecting part of the correlations between different variables. Such an approximation is called the mean field approximation.

Flory Exponent

In the derivation of the Flory exponent (2.56) using the mean field approximation, the following two important contributions originating from the fluctuations and correlations cancel each other out:

- There is an overestimate of the conformation entropy due to the use of the ideal chain statistics given by (2.54). Indeed, the number of possible conformations of an excluded volume chain is much less than that of an ideal chain because of the constraint on the spatial arrangements of the segments. In such a case, the conformation entropy of the excluded volume chain is much less than that of the ideal chain.
- There is an overestimate of the interaction energy due to the mean field approximation (2.55). In general, there is a large inhomogeneity in the segment distribution inside the coil of the chain. Thus contacts between segments take place mainly in the high-density region of the segments. The approximation using the average segment density (2.55) overestimates the total number of contacts between segments.

A calculation that takes the effects of the fluctuations in the segment density into account has been done using the renormalization group theory. This gives an exponent value of 0.588, very close to the Flory value 3/5 [14].

The simplest example of a mean field approximation for polymer chains is Flory's theory of an isolated excluded volume chain. When we consider a linear polymer chain made of N segments $(N \gg 1)$, the total free energy of this chain F is given as a sum of the conformation entropy F_{conf} and the segment interaction F_{int} as

$$F = F_{\text{conf}} + F_{\text{int}} \,. \tag{2.53}$$

We use the conformation entropy of an ideal chain whose end-to-end distance is R. From (2.37), we then obtain

$$F_{\text{conf}}(R) = \frac{3k_{\text{B}}TR^2}{2Nb^2} \,. \tag{2.54}$$

On the other hand, the excluded volume interaction is evaluated by assuming that N segments are distributed uniformly within a sphere of radius R. In the mean field approximation, we assume that the number of contacts between segments is proportional to the square of the segment density. Using the excluded volume parameter v introduced in (2.52), the interaction energy is evaluated within such an approximation as

$$F_{\text{int}} = \frac{v}{2} \int_{\substack{\text{sphere of}\\ \text{radius } R}} (\text{average segment density})^2 \mathrm{d}\boldsymbol{r} \sim \frac{3vN^2}{8\pi R^3} \,. \tag{2.55}$$

Substituting (2.54) and (2.55) into (2.53) and minimizing the resulting free energy with respect to the radius R, we obtain the equilibrium radius[8]

$$R_{\text{MF}} = \left(\frac{3vb^2N^3}{8\pi k_{\text{B}}T} \right)^{1/5} \sim N^{3/5} \,, \tag{2.56}$$

where the subscript MF stands for the mean field approximation.

Renormalization Group Theory

The renormalization group theory [14] for polymers is a method in which one repeats a coarse-graining of the chain by grouping several segments into a coarse-grained segment and rescaling the unit of length. The important model parameters for isolated excluded volume chains are the segment size b, the total number of segments in the chain N, and the excluded volume parameter v. When we define a coarse-grained segment by grouping m segments, the above model parameters change as $N \to N/m$, $b \to m^{\nu}b$, and $v \to m^{\nu'}v$, where ν and ν' are constants.

When such a coarse-graining procedure is applied, the unit of length is redefined (rescaled) so that the size of the unit of the chain is unchanged before and after the transformation. The combined coarse-graining and rescaling procedure is the renormalization transformation (see Fig. 2.22). The map relating the parameters before and after the transformation is determined in such a way that the canonical probability distributions of the degrees of freedom surviving after the transformation are invariant. Repeating the renormalization transformation, we finally reach a state where the model parameters no longer change. Such a state is called a fixed point. At the fixed point, the statistical properties of the chain are preserved after coarsening and rescaling. The structure thus has the self-similar nature discussed in the box entitled *Ideal Chains and Fractal Structure* on p. 41. The ideal Gaussian chain with the scaling law $R \sim N^{1/2}$ and the excluded volume chain with the scaling law $R \sim N^{3/5}$ are known to correspond to such fixed points.

The exponent 3/5 is larger than the value 1/2 for the ideal chain given in (2.9), which means that the excluded volume chain is swollen. This exponent

[8] In the d-dimensional space, the exponent becomes $3/(d+2)$.

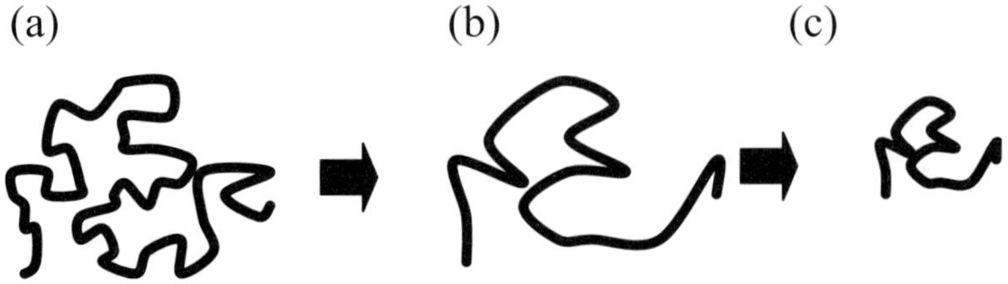

Fig. 2.22. Schematic illustration of the renormalization transformation for a polymer chain. (**a**) The original chain. (**b**) The coarse-grained chain, where a coarse-grained segment is defined by grouping several original segments. (**c**) The unit of length is rescaled. At a fixed point of this renormalization transformation, the statistical properties of the chain after the transformation [chain (**c**)] coincide with those of the chain before the transformation [chain (**a**)]. The chain conformation at a fixed point is thus a self-similar fractal structure

3/5 is called the Flory exponent for excluded volume chains. It is known that the Flory exponent agrees quite well with the experimental result in spite of the rough approximations used in the theory (see Fig. 2.12). However, this does not mean that the mean field theory is accurate. In fact, the success of (2.56) is due to a cancellation between the overestimate of the conformation entropy in (2.54) and the overestimate of the interaction energy in (2.55) (see the box entitled *Flory Exponent* on p. 50).

In a dilute polymer solution, the segment density fluctuates considerably from point to point. In such a system, the correlation effects in the segment density are so large[9] that the mean field approximation given in (2.55), which neglects the fluctuation, cannot be justified. We must then ask in which situation the mean field approximation becomes valid.

Such a situation is realized in concentrated polymer systems where the chains are strongly overlapping and fluctuations in the segment density are suppressed. The repulsive excluded volume interaction acting on a tagged segment is composed of two types of contribution: one is the contribution from segments that belong to the same chain as the tagged segment, and the other is the contribution from segments that belong to the other chains. In a concentrated polymer solution or in a polymer melt, the sum of these two contributions should cancel each other because of the incompressibility condition (see the box entitled *Screening Effects in Concentrated Polymer Systems* in on p. 55.) This leads to an ideal chain behavior of a polymer chain in a concentrated system. Such an effect is called the screening effect.

[9] For example, when we focus on a certain segment, we can observe that the surrounding segments are attracted to or repelled from this segment due to the interaction between segments. Such an interaction produces fluctuations in the segment density around the uniform distribution.

Due to this screening effect, the mean field approximation using ideal chain statistics becomes effective in concentrated polymer systems. Many successful examples of the mean field theory will be given in Chaps. 3 and 4.

Perturbation Theory

In this section, we discuss the second approximate technique: perturbation theory. As an example, we consider the radius of gyration of an isolated excluded volume chain. As shown in (2.30), the Hamiltonian of an excluded volume chain is given by the sum of the ideal chain part $\mathcal{H}_0$ and an interaction part $\mathcal{W}$ as follows:

$$\mathcal{H} = \mathcal{H}_0 + \mathcal{W} . \tag{2.57}$$

In the perturbation theory, the interaction part $\mathcal{W}$ is assumed to be much smaller than the thermal fluctuation energy $k_{\mathrm{B}}T$. Then the canonical probability distribution function of the chain conformation is expanded as

$$\exp\Big[-\beta(\mathcal{H}_0 + \mathcal{W})\Big] = \Big[1 - \beta\mathcal{W} + \frac{1}{2}(\beta\mathcal{W})^2 + \cdots\Big]\exp\Big(-\beta\mathcal{H}_0\Big) . \tag{2.58}$$

Using this relation, the average $\langle\ \rangle$ over the canonical distribution of the Hamiltonian $\mathcal{H}$ can be rewritten in terms of the average over the canonical distribution of the ideal chain Hamiltonian $\langle\ \rangle_0$ as

$$\begin{aligned}\langle * \rangle \equiv \frac{\displaystyle\int * \, \mathrm{e}^{-\beta\mathcal{H}}\mathrm{d}\Gamma}{\displaystyle\int \mathrm{e}^{-\beta\mathcal{H}}\mathrm{d}\Gamma} &= \frac{\displaystyle\int * \, (1 - \beta\mathcal{W} + \cdots)\mathrm{e}^{-\beta\mathcal{H}_0}\mathrm{d}\Gamma}{\displaystyle\int (1 - \beta\mathcal{W} + \cdots)\mathrm{e}^{-\beta\mathcal{H}_0}\mathrm{d}\Gamma} \\ &= \frac{\langle * \, (1 - \beta\mathcal{W} + \cdots)\rangle_0}{\langle 1 - \beta\mathcal{W} + \cdots\rangle_0} ,\end{aligned} \tag{2.59}$$

where $\int \mathrm{d}\Gamma = \prod_i \int \mathrm{d}\boldsymbol{r}_i$ is an integration over the whole phase space spanned by the segment positions $\{\boldsymbol{r}_i\}$. Using this expansion, the canonical average $\langle\ \rangle$ of the system with the segment interaction can be rewritten using the average $\langle\ \rangle_0$ over the Gaussian distribution $\exp(-\beta\mathcal{H}_0)$ of an ideal chain, i.e., Gaussian integrals that can be performed analytically.

In the perturbation theory for the end-to-end distance R of an excluded volume chain, the relation (2.59) applied to $\langle|\boldsymbol{r}_N - \boldsymbol{r}_0|^2\rangle$ leads to [15]

$$\langle|\boldsymbol{r}_N - \boldsymbol{r}_0|^2\rangle = Nb^2\left(1 + \frac{4}{3}z - 2.075z^2 + \cdots\right) , \tag{2.60}$$

where the smallness parameter z is given by

$$z \equiv \left(\frac{3}{2\pi}\right)^{3/2}\frac{vN^{1/2}}{b^3} . \tag{2.61}$$

Because this smallness parameter z is proportional to $vN^{1/2}$, even if the excluded volume parameter is very small ($v/b^3 \ll 1$), the perturbation expansion given in (2.60) breaks down in the limit of long chain length N. Thus the applicability of the perturbation theory to dilute polymer solutions is very limited. On the other hand, in concentrated polymer systems, perturbation theory can give us important information for a qualitative understanding of the phenomena. A popular example of perturbation theory in realistic applications is the random phase approximation in which a small deformation in the polymer conformation is linearized around the ideal chain conformation. Examples of this technique will be given in Sect. 4.1.3.

2.3.4 Statistical Properties of Many-Chain Systems

Dilute and Concentrated Polymer Solutions

We have been discussing the statistical properties of a single polymer chain. We now move on to the essential properties of many-chain systems. A polymer solution is a mixture of polymers and solvents, the latter being in general composed of low-molecular-weight components. As shown in Fig. 2.23, polymer solutions are categorized into several classes. In the dilute solution regime, polymer chains are isolated in the matrix of the solvent. As we increase the density of the polymer chains, the coils of the chains start to contact each other at a certain polymer density. Such a density is called the overlapping density, which is the boundary between the dilute regime and the concentrated regime. Increasing the polymer density beyond the overlapping density, the coils of the chains start to overlap and the polymer chains form a network structure, while the volume fraction of polymers in the total volume of the system is still very low. Such a regime is called the semi-dilute solution regime. The existence of the semi-dilute regime is characteristic of polymer solutions.

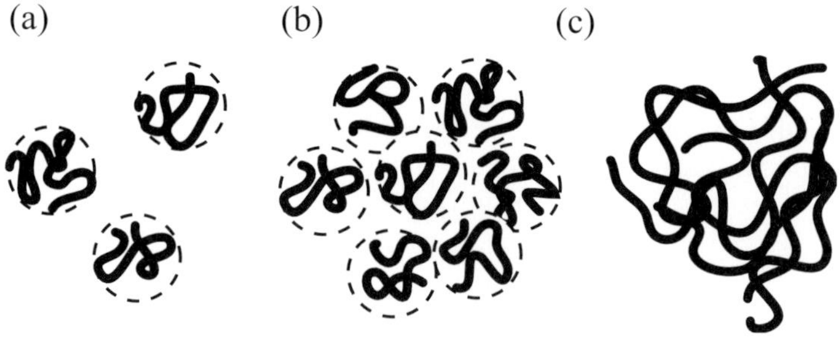

Fig. 2.23. Several categories of polymer solutions. (**a**) The dilute regime, (**b**) the overlapping density, and (**c**) the semi-dilute and concentrated regimes

Further increasing the polymer density will bring the solution through the concentrated solution regime until finally it reaches the polymer melt regime, which is a liquid state of polymers without solvents.

Flory–Huggins Free Energy

In a polymer melt, the excluded volume interaction between segments is screened due to the incompressibility condition and overlapping between chains. In such a case, each polymer chain in the melt obeys ideal chain statistics (see the box entitled *Screening Effects in Concentrated Polymer Systems*). In multicomponent polymer melts, there exists another type of segment interaction that is not completely screened. A typical example is a binary melt of A and B polymers. Let us assume for simplicity that the volumes of the A and B segments are the same, and use the lattice model introduced in Sect. 2.1.1. If we assume that the chains are the excluded volume chains shown in Fig. 2.5b, each lattice site is occupied by either an A or a B segment.

Screening Effects in Concentrated Polymer Systems

In semi-dilute polymer solutions and polymer melts, the excluded volume interaction between segments effectively vanishes, and the chain obeys ideal chain statistics. This originates from the suppression of segment density fluctuations due to the very low compressibility (almost incompressible) of polymer melts. Such an effect is called the screening effect.

Let us consider the conformation of a single (tagged) chain in a concentrated polymer solution or a polymer melt composed of a single type of chain. We pick out one of the segments of this chain and call it the tagged segment. There are two types of excluded volume interaction acting on this tagged segment:

1. The interaction between the tagged segment and other segments belonging to the tagged chain.
2. The interaction between the tagged segment and segments that belong to other chains than the tagged chain.

As the system is composed of the same type of chain, the tagged chain cannot distinguish (1) and (2). Due to the incompressibility condition, the sum of the two interactions (1) and (2) should be almost zero everywhere because the total segment density is constant. This means that the driving force for the tagged chain to swell [interaction (1)] is almost exactly cancelled by the pressure from the other chains [interaction (2)]. Hence the excluded volume interaction is screened.

The equilibrium state of a multicomponent polymer melt is the state that minimizes the Helmholtz free energy

$$F = E - TS \; , \tag{2.62}$$

where E is the interaction energy between the segments and S is a sum of the conformational entropy and the translational entropy of the centers of mass of the chains (the so-called mixing entropy).

Let us first evaluate the interaction energy E using the mean field approximation. For simplicity, we assume that there exist only pair interactions between pairs of segments on nearest neighbor lattice sites. We introduce an index l to specify the lattice site and denote the position vector of the lth lattice site by $\boldsymbol{r}_l$. We then denote the average number of K-type segments on the lth lattice site by $\phi_K(\boldsymbol{r}_l)$ $(0 \le \phi(\boldsymbol{r}_l) \le 1)$, and call it the segment volume fraction. (The term 'volume fraction' means the ratio of the volume occupied by the target to the total volume of the space.) The quantity $\phi_K(\boldsymbol{r}_l)$ corresponds to the probability of finding a K-type segment at the lattice site labeled l.

Let us denote the interaction energy of A–A, B–B, and A–B pairs on nearest neighbor lattice sites by ϵ_{AA}, ϵ_{BB}, and ϵ_{AB} (see Fig. 2.24). The total interaction energy of the system is then written under the mean field approximation as

$$E = \frac{1}{2} \sum_l \sum_{\boldsymbol{a}} \sum_{K=A,B} \sum_{K'=A,B} \epsilon_{KK'} \phi_K(\boldsymbol{r}_l) \phi_{K'}(\boldsymbol{r}_l + \boldsymbol{a}) \; , \tag{2.63}$$

where $\sum_{\boldsymbol{a}}$ is a sum over the z vectors $\boldsymbol{a}$ that point to the nearest neighbor lattice sites, as introduced in (2.21). Approximating the position-dependent segment density $\phi_K(\boldsymbol{r}_l)$ by the mean value over the system $\bar{\phi}_K$, we obtain

$$E = \frac{z N_{\mathrm{L}}}{2} \sum_{KK'} \epsilon_{KK'} \bar{\phi}_K \bar{\phi}_{K'} \; , \tag{2.64}$$

where N_{L} is the total number of lattice sites in the system.

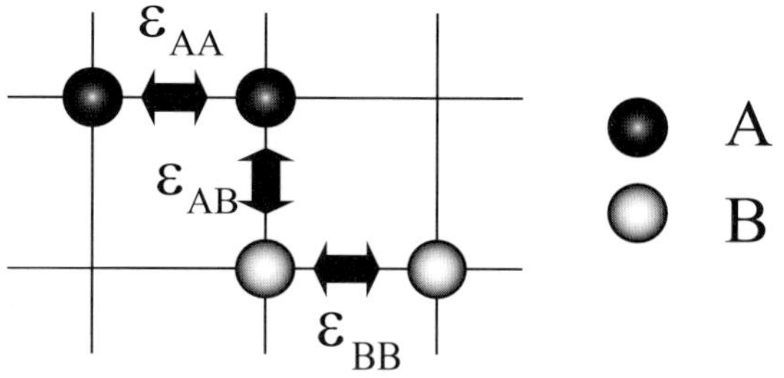

Fig. 2.24. Definition of the segment interaction parameter $\epsilon_{KK'}$ in the lattice model

Next we consider the entropic contribution S. Due to the screening effect, the conformation of an individual chain is close to that of the ideal chain.[10] Therefore, we can assume that the conformation entropy of the chains is invariant, even if the system is inhomogeneous. In such a case, the dominant contribution to the entropy term S comes from the translational degrees of freedom of the center of mass of the individual chain. If the system is homogeneous, we can use the expression for the mixing entropy of an ideal gas given by

$$\begin{aligned} S &= k_{\mathrm{B}} \ln \begin{pmatrix} \text{Number of possible arrangements} \\ \text{of centers of mass of chains} \end{pmatrix} \\ &= k_{\mathrm{B}} \ln \frac{(M_A + M_B)!}{M_A! M_B!} \\ &\sim -k_{\mathrm{B}} N_{\mathrm{L}} \left(\frac{\bar{\phi}_A}{N_A} \ln \bar{\phi}_A + \frac{\bar{\phi}_B}{N_B} \ln \bar{\phi}_B \right) , \end{aligned} \tag{2.65}$$

where N_K is the total number of segments in a K-type chain and M_K is the total number of K-type chains in the system. When we derive the third line from the second line, we use Stirling's formula given by

$$\ln N! \sim N \ln N - N \quad \text{for } N \gg 1 , \tag{2.66}$$

and neglect constant terms and terms linear in $\bar{\phi}_K$. The terms linear in $\bar{\phi}_K$ do not affect the phase behavior of the mixture because such linear terms produce an extra constant in the chemical potential due to the relation $\mu_K = \partial F / \partial \bar{\phi}_K$, which is cancelled by the equilibrium condition $\mu_K = \mu_{K'}$ for all K and K' pairs. Equation (2.65) is the increase in the entropy upon mixing different types of molecules, and is called the mixing entropy (see the box entitled *Mixing Entropy*).

Mixing Entropy

The mixing entropy in (2.65) accounts for the increase in the entropy when different types of molecules are mixed. This corresponds to the increase in the possible numbers of arrangements of the centers of mass of the molecules upon mixing, and gives the thermodynamic driving force for two different materials to mix together. In the polymer case, the mixing entropy is determined only by the arrangements of the centers of mass of the chains and is not dependent on the conformational degrees of freedom.

The miscibility of two types of polymer is determined by competition between the mixing entropy and the segment interaction. In polymeric systems,

[10] Within the mean field approximation, deviation of the chain conformation from the ideal chain conformation occurs only when there is inhomogeneity in the segment densities, such as a chain in a phase-separating polymer blend or a chain near a solid surface.

the total number of translational degrees of freedom of the chains is much lower than the total number of segments. Therefore, the effect of the mixing entropy is extremely small. This is why a mixture of polymers easily goes into phase separation due to a small difference in the chemical nature of the segments. For example, a mixture of protonated polyethylene and deuterated polyethylene separates into two phases when the molecular weights are very large, although the chemical properties of their monomers are almost identical.

Combining (2.64) and (2.65), we can obtain the expression for the dimensionless free energy, i.e., the free energy divided by $k_{\mathrm{B}}T$, per lattice site as

$$\frac{F}{N_{\mathrm{L}} k_{\mathrm{B}} T} = \sum_K \frac{\bar{\phi}_K}{N_K} \ln \bar{\phi}_K + \frac{1}{2} \sum_{KK'} \chi_{KK'} \bar{\phi}_K \bar{\phi}_{K'} , \tag{2.67}$$

where $\chi_{KK'}$ is the segment interaction parameter defined by[11]

$$\chi_{KK'} = \frac{z\epsilon_{KK'}}{k_{\mathrm{B}}T} . \tag{2.68}$$

The free energy given by (2.67) is called the Flory–Huggins free energy, where only the segment interaction and the mixing entropy are taken into account. Although the conformation entropy is also taken into account in the standard derivation of the Flory–Huggins theory, the chain conformation is always an ideal Gaussian one, irrespective of the composition of the system, because of our assumption of spatial uniformity and the screening effect. This means that the conformation entropy of an individual chain is always constant as long as the system is uniform. For this reason, the contribution from the chain conformation entropy to the free energy is linear in the segment density and is cancelled in the equilibrium condition. In Chap. 3, the effect of the conformation entropy will be correctly taken into account using the self-consistent field calculation.

We now consider a polymer solution where one of two components A and B is a polymeric component and the other is a low-molecular-weight solvent. Specifying the segment of the polymer and the solvent by indices P and S, (2.67) is rewritten as

$$\frac{F}{N_{\mathrm{L}} k_{\mathrm{B}} T} = \frac{\bar{\phi}_{\mathrm{P}}}{N_{\mathrm{P}}} \ln \bar{\phi}_{\mathrm{P}} + \bar{\phi}_{\mathrm{S}} \ln \bar{\phi}_{\mathrm{S}} + \frac{1}{2} \left(\chi_{\mathrm{PP}} \bar{\phi}_{\mathrm{P}}^2 + \chi_{\mathrm{SS}} \bar{\phi}_{\mathrm{S}}^2 + 2\chi_{\mathrm{PS}} \bar{\phi}_{\mathrm{P}} \bar{\phi}_{\mathrm{S}} \right) , \tag{2.69}$$

where we have assumed that the volume of a single solvent molecule is the same as that of a single segment. A comparison between the theoretical phase diagram of a polymer solution obtained using (2.69) and the experimental

[11] If we do not use the lattice model, z corresponds to the average number of segments around the central segment (average coordination number).

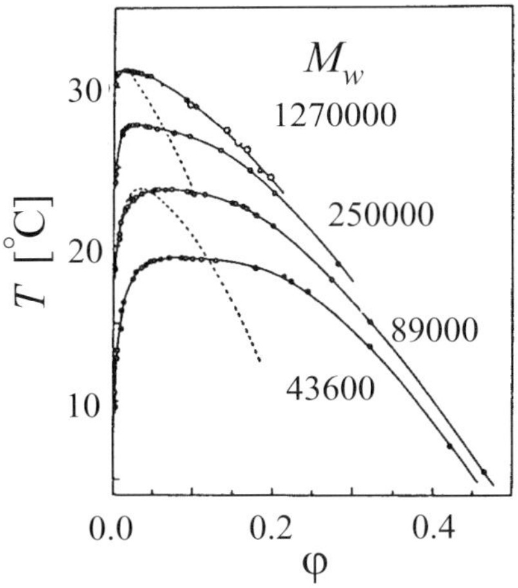

Fig. 2.25. Experimentally determined phase diagram of a polystyrene dissolved in a cyclohexane solvent (*symbols* and *continuous curves*) and a theoretical phase diagram calculated using the Flory–Huggins free energy model (*broken curve*). Reprinted with permission from A.R. Schultz and P.J. Flory, J. Am. Chem. Soc. **74**, 4760 (1952). (© 1952, American Chemical Society)

phase diagram is shown in Fig. 2.25. In polymer solutions, due to the large difference in the molecular weights between the polymeric component and the solvent, the phase diagram is highly asymmetric with respect to the composition, and the polymer solution shows a phase separation between a polymer-rich phase and a solvent-rich phase, even in a very dilute region of the polymer component. This means that it is difficult to dissolve a polymer with a large molecular weight in a solvent and the polymers tend to form precipitates.

Free Energy and Excluded Volume Parameter of Polymer Solutions

Let us evaluate the effective two-body interaction between segments by eliminating the solvent degrees of freedom in (2.69). We first substitute the relation $\bar{\phi}_{\mathrm{S}} = 1 - \bar{\phi}_{\mathrm{P}}$ obtained from the incompressibility condition into (2.69) to eliminate $\bar{\phi}_{\mathrm{S}}$. Then we assume that the polymer density $\bar{\phi}_{\mathrm{P}}$ is small enough to ensure that the system is in the dilute solution regime. Using the approximation $\ln(1 - \bar{\phi}_{\mathrm{P}}) \sim -\bar{\phi}_{\mathrm{P}} - \bar{\phi}_{\mathrm{P}}^2/2$, one obtains the term containing $\bar{\phi}_{\mathrm{P}}^2$ as

$$\frac{1}{2}\left\{1 - 2\left[\chi_{\mathrm{PS}} - \frac{1}{2}(\chi_{\mathrm{PP}} + \chi_{\mathrm{SS}})\right]\right\}\bar{\phi}_{\mathrm{P}}^2 \,. \tag{2.70}$$

As the coefficient of $\bar{\phi}_{\mathrm{P}}^2$ corresponds to half of the pair interaction energy between two segments under the influence of the solvent molecules, it can be identified with the excluded volume parameter v introduced in Sect. 2.1.1. Rewriting the volume fraction $\bar{\phi}_{\mathrm{P}}$ in (2.70) with the number density, an extra factor b_0^3 appears from the Jacobian, and the expression for the excluded volume parameter can be given as

$$v = b_0^3 \left\{ 1 - 2 \left[\chi_{\mathrm{PS}} - \frac{1}{2}(\chi_{\mathrm{PP}} + \chi_{\mathrm{SS}}) \right] \right\} . \tag{2.71}$$

The first term in curly brackets on the right-hand side of (2.71) comes from the mixing entropy, and the second term containing $\chi_{KK'}$ terms is the contribution from the interaction between segments. For an A/B binary mixture, the parameter defined by

$$\chi \equiv \chi_{AB} - \frac{1}{2}(\chi_{AA} + \chi_{BB}) \tag{2.72}$$

is called the χ-parameter (see the box entitled *χ-Parameter*). This χ-parameter is a measure of the relative strength of the repulsion between different types of segment to the attraction between the same types of segment. The smaller the value of χ, the larger the miscibility between A and B. Applying the definition of (2.72) to (2.71), we find the relation

$$v = b_0^3(1 - 2\chi) . \tag{2.73}$$

Volume Fraction, Segment Number Density and Excluded Volume Parameter

The volume fraction ϕ_K is dimensionless because it is a ratio between the volume occupied by K-type segments and the total volume. On the other hand, the number density of the K-type segments, denoted n_K, is defined as the number of segments per unit volume, and has the dimension of an inverse volume. Hence, the relation between the average volume fraction of the polymer segments, $\bar{\phi}_{\mathrm{P}}$, and the average number density of the polymer segments, $\bar{n}_{\mathrm{P}}$, is given by

$$\bar{\phi}_{\mathrm{P}} = b_0^3 \bar{n}_{\mathrm{P}} . \tag{2.74}$$

Substituting this relation into (2.69), we obtain

$$\begin{aligned} \frac{F}{\mathcal{V} k_{\mathrm{B}} T} &= \frac{\bar{n}_{\mathrm{P}}}{N_{\mathrm{P}}} \ln \bar{n}_{\mathrm{P}} + \frac{1}{2} b_0^3 \left[1 - 2\chi_{\mathrm{PS}} - \frac{1}{2}(\chi_{\mathrm{PP}} + \chi_{\mathrm{SS}}) \right] \bar{n}_{\mathrm{P}}^2 \\ &\quad + (0\,\text{th and } 1\,\text{st order terms in } \bar{n}_{\mathrm{P}}) . \end{aligned} \tag{2.75}$$

The excluded volume parameter v is defined as the prefactor of the $\bar{n}_{\mathrm{P}}^2$ term, which leads to the definition of (2.71).

The derivation of (2.73) from (2.69) indicates that the excluded volume parameter v is determined by competition between two contributions. The first term in brackets on the right-hand side of (2.73) corresponds to the average free volume assigned to each segment, which depends only weakly on the temperature. On the other hand, the second term accounts for the difference in the A–A, A–B and B–B type pair interactions, which is inversely proportional to the temperature, as explained in (2.68). Because of these different temperature dependences, the excluded volume parameter v changes sign from negative to positive when the temperature is raised. This is the reason underlying the coil–globule transition shown in Fig. 2.8.

χ-Parameters

In lattice models, it is assumed that the segment has no internal structure and that the distance between the interacting segments is always constant. In this case, it is reasonable to assume that the interaction parameter $\epsilon_{KK'}$ in (2.68) has no temperature dependence. Using this assumption and (2.72), we can immediately conclude that the χ-parameter has the temperature dependence

$$\chi = \frac{A}{k_{\mathrm{B}}T} ,$$

where A is a constant. However, as an actual segment is composed of many monomers, it has a complex internal structure. Thus the entropic contribution from such internal degrees of freedom of the segment cannot be neglected. If we denote the number of possible configurations of the internal degrees of freedom by W, the entropic contribution from the internal degrees of freedom to the segment interaction parameter is given by

$$-TS = -k_{\mathrm{B}}T \ln W .$$

If the number of internal configurations W does not strongly depend on the temperature, this effect gives a constant contribution to the χ-parameter, which is now written as

$$\chi = \frac{A}{k_{\mathrm{B}}T} + B ,$$

where A and B are constants. The temperature dependence of real χ-parameters is much more complex, and many empirical laws for the temperature dependence of the χ-parameter have been proposed [16].

2.4 Dynamical Models of a Polymer Chain Based on a Molecular Description

Up to now, we have been discussing the equilibrium properties of polymer chains. In this section, we introduce several simple dynamical models and show their applications.

In polymer solutions or melts, a chain moves randomly due to the fluctuating forces imposed by the solvent molecules or other chains (see Fig. 2.26). Such random motion is called Brownian motion.

Because the size of a conventional polymer chain lies beyond the resolution of optical microscopes, it is impossible to perform an in situ observation of the motion of a single chain. However, using the fluorescent microscopy technique and biological polymers such as DNA molecules or actin filaments, one can observe such chain motion directly (see Sect. 2.6). The direct observation technique shows the strong influence of the fluctuating force on the chain dynamics. In this section, we develop a dynamical model for the Brownian motion of an isolated chain and study its behavior using several theoretical methods.

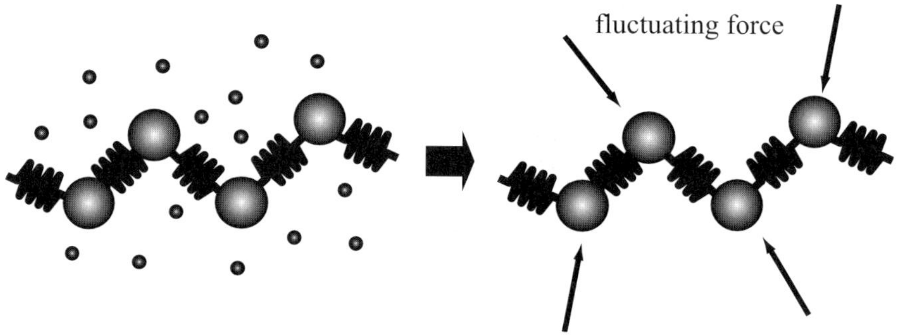

Fig. 2.26. Motion of a polymer chain driven by the random fluctuating forces imposed by solvent molecules

2.4.1 Formulation of Brownian Motion

Langevin Equation and Thermal Fluctuation

Before going into the details of models for the Brownian motion of polymer chains, we briefly summarize the general properties of Brownian motion.

There are two standard descriptions of Brownian motion. One is the Langevin equation and the other is the Fokker–Planck equation. The Langevin equation is a stochastic differential equation describing the time evolution of the random variables under the influence of an external random force. On the other hand, the Fokker–Planck equation is a partial differential equation describing the time evolution of the probability distribution function of the random variable. As these two descriptions are equivalent for a certain class of random processes, we will use the Langevin equation in the following (see the box entitled *Langevin Equation and Fokker–Planck Equation*).

Let us consider the one-dimensional motion of a single Brownian particle in a solvent. This Brownian particle is subjected to the random fluctuating force imposed by the solvent molecules. If we denote the position of the Brownian particle at time t by $x(t)$, this $x(t)$ can be regarded as a random variable whose time evolution is given by

$$m\frac{\mathrm{d}^2}{\mathrm{d}t^2}x(t) = -\zeta\frac{\mathrm{d}}{\mathrm{d}t}x(t) - \frac{\mathrm{d}}{\mathrm{d}x}U(x) + \xi(t) , \tag{2.76}$$

where m is the mass of the Brownian particle, ζ is the friction constant of the Brownian particle in the solvent, $-\mathrm{d}U(x)/\mathrm{d}x$ is the force due to the potential $U(x)$, and $\xi(t)$ is the random fluctuating force imposed by the solvent.

We now assume that the solvent is so viscous ($(\zeta/m \gg 1)$) that the inertia term in the equation of motion can be neglected. The Brownian particle is therefore assumed to move with the terminal velocity determined by the balance between the friction force and the external force acting on the particle. In such a limiting situation, (2.76) reduces to

$$\zeta\frac{\mathrm{d}}{\mathrm{d}t}x(t) = -\frac{\mathrm{d}}{\mathrm{d}x}U(x) + \xi(t) . \tag{2.77}$$

The origin of the fluctuating force $\xi(t)$ is a huge number of independent collisions between the Brownian particle and the solvent molecules. Due to the isotropy of the solvent, the mean value of the fluctuating force is assumed to be zero. The independence between different collisions with the solvent molecules means that the correlation between random forces at different times vanishes. We thus have the relation

$$\langle\xi(t)\rangle = 0 , \quad \langle\xi(t)\xi(t')\rangle = 2\alpha\delta(t - t') , \tag{2.78}$$

where the coefficient 2α is the variance of $\xi(t)$. As (2.77) contains a random force, it is called a stochastic differential equation or a Langevin equation (see the box entitled *Langevin Equation and Fokker–Planck Equation*).

Langevin Equation and Fokker–Planck Equation

Let us consider the following Langevin equation, which has the same form as (2.77):

$$\frac{\mathrm{d}}{\mathrm{d}t}x(t) = f(x(t)) + \xi(t) ,$$

$$\langle\xi(t)\rangle = 0 , \quad \langle\xi(t)\xi(t')\rangle = 2\alpha\delta(t - t') ,$$

where $x(t)$ is a random variable, $f(x)$ is the average rate of change of x, $\xi(t)$ is a random fluctuating force, and α is the variance of ξ.

By generating many samples of $x(t)$ that obey the above Langevin equation, we can define the probability distribution function of the random variable x at time t. Let us denote such a probability distribution function by $P(x,t)$. If the random force $\xi(t)$ obeys the Gaussian distribution, the following evolution equation for $P(x,t)$ is derived from the Langevin equation [17]:

$$\frac{\partial}{\partial t}P(x,t) = -\frac{\partial}{\partial x}\left[f(x)P(x,t)\right] + \alpha\frac{\partial^2}{\partial x^2}P(x,t) .$$

This equation is called the Fokker–Planck equation. As long as we are concerned with Gaussian random forces, the Langevin equation and the Fokker–Planck equation are equivalent. We can derive one from the other.

Fluctuation–Dissipation Theorem

One of the important theorems in non-equilibrium statistical mechanics is the so-called fluctuation–dissipation theorem which relates the coefficient α in the correlation of the random force given by (2.78) to the friction coefficient ζ according to

$$\alpha = \zeta k_{\mathrm{B}} T . \tag{2.79}$$

Fluctuation–Dissipation Theorem

Let us consider a random variable $x(t)$ that obeys the Langevin equation defined by (2.77) and (2.78). As described in the box entitled *Langevin Equation and Fokker–Planck Equation*, the Fokker–Planck equation corresponding to this Langevin equation is

$$\frac{\partial}{\partial t}P(x,t) = -\frac{\partial}{\partial x}\left\{\left[-\frac{1}{\zeta}\frac{\mathrm{d}U(x)}{\mathrm{d}x} - \frac{\alpha}{\zeta^2}\frac{\partial}{\partial x}\right]P(x,t)\right\} .$$

As this equation has the form of the continuity equation in hydrodynamics, the quantity between curly brackets on the right-hand side is the flux of the probability distribution $P(x,t)$. Because the probability flux should vanish in the equilibrium state, we obtain

$$\left(-\frac{1}{\zeta}\frac{\mathrm{d}U}{\mathrm{d}x} - \frac{\alpha}{\zeta^2}\frac{\partial}{\partial x}\right)P(x,t) = 0 .$$

The solution of this differential equation gives the equilibrium distribution of the random variable x. It should be identical to the canonical distribution

$$P_{\mathrm{eq}}(x) = C\exp\left[-\beta U(x)\right] ,$$

where C is an integration constant. Such a requirement leads to the fluctuation–dissipation relation (2.79).

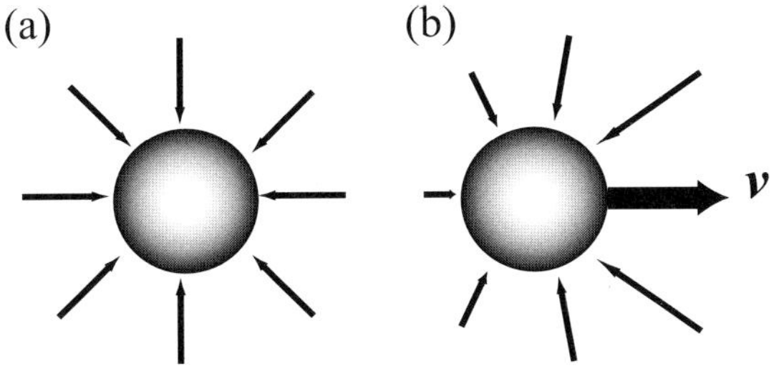

Fig. 2.27. Friction force and fluctuating force acting on a Brownian particle in a solvent. (**a**) If the Brownian particle is at rest, the fluctuating force is isotropic. (**b**) On the other hand, if the Brownian particle is moving with a non-zero mean velocity, the random force from the front side becomes larger than that from the rear side. This difference produces the friction force

As shown in Fig. 2.27, when the Brownian particle is moving with non-zero mean velocity, collisions with solvent molecules from the front side are more frequent than those from the rear side. The excess collisions from the front side are the origin of the friction force. We can thus understand that the friction force and the fluctuating force have the same origin, which leads to the fluctuation–dissipation relation given in (2.79) (see the box entitled *Fluctuation–Dissipation Theorem*).

Einstein's Relation

In the case where there is no potential term $U(x)$ in (2.77), let us prove the relation (2.83). Setting $U = 0$ in (2.77) and integrating with respect to time, we obtain

$$x(t) - x(0) = \frac{1}{\zeta}\int_0^t \mathrm{d}s\,\xi(s)\,. \tag{2.80}$$

Using this relation, the mean square displacement of the Brownian particle is calculated as

$$\begin{aligned}\left\langle [x(t) - x(0)]^2 \right\rangle &= \frac{1}{\zeta^2}\int_0^t \mathrm{d}s \int_0^t \mathrm{d}s' \langle \xi(s)\xi(s')\rangle \\ &= \frac{1}{\zeta^2}\int_0^t \mathrm{d}s \int_0^t \mathrm{d}s' 2\alpha\delta(s-s') = \frac{2\alpha t}{\zeta^2}\,.\end{aligned} \tag{2.81}$$

As the left-hand side of this equation is $2Dt$ according to the definition of the diffusion coefficient for one-dimensional Brownian motion [see (2.2)], the relation (2.83) can be obtained immediately. Combining this relation with the fluctuation–dissipation relation (2.79), we obtain Einstein's relation

$$D = \frac{k_{\mathrm{B}}T}{\zeta}\,. \tag{2.82}$$

Another important property associated with the fluctuating force is derived when there is no potential $U(x)$. In this case, the parameter α in (2.78) is related to the diffusion constant D of a one-dimensional random walk defined in (2.2) by (see the box entitled *Einstein's Relation*)

$$\alpha = D\zeta^2 . \tag{2.83}$$

Combining this relation and the fluctuation–dissipation relation given by (2.79), we obtain

$$D = \frac{k_{\mathrm{B}}T}{\zeta} . \tag{2.84}$$

This relation is called Einstein's relation. Equation (2.84) is consistent with our intuition that the more viscous the solvent is, the less easily the Brownian particle can diffuse.

2.4.2 Rouse Model of a Single Polymer Chain in a Solvent

General Equation of Motion for a Chain in a Flow

The simplest model for the Brownian motion of a polymer chain can be obtained by combining the bead–spring model introduced in Sect. 2.2 and the Brownian motion of each segment.

The individual segment is driven by the sum of three forces, i.e., the potential force derived from the Hamiltonian of an ideal chain (2.31), the friction force proportional to the velocity difference between the segment and the surrounding solvent, and the random fluctuating force. Hence, the equation of motion for the i th segment is given by

$$\zeta \left[\frac{\mathrm{d}\boldsymbol{r}_i(t)}{\mathrm{d}t} - \boldsymbol{v}(\boldsymbol{r}_i) \right] = k_0 \left(\boldsymbol{r}_{i+1} + \boldsymbol{r}_{i-1} - 2\boldsymbol{r}_i \right) + \boldsymbol{\xi}_i(t) , \tag{2.85}$$

where $k_0 = 3k_{\mathrm{B}}T/b^2$ is the spring constant of the bond connecting the segments [see (2.28)] and $\boldsymbol{v}(\boldsymbol{r})$ is the flow velocity of the solvent at position $\boldsymbol{r}$. To treat the end segments in the same manner as the other segments, we introduce fictitious segments $\boldsymbol{r}_{-1}$ and $\boldsymbol{r}_{N+1}$ and assume the conditions

$$\boldsymbol{r}_{-1} = \boldsymbol{r}_0 \quad \text{and} \quad \boldsymbol{r}_{N+1} = \boldsymbol{r}_N . \tag{2.86}$$

The last term $\boldsymbol{\xi}_i(t)$ on the right-hand side of (2.85) is the fluctuating force acting on the i th segment at time t. Applying (2.78) and (2.79) to $\boldsymbol{\xi}_i(t)$, we obtain the fluctuation–dissipation relation

$$\left\langle \boldsymbol{\xi}_i(t)\boldsymbol{\xi}_j(t') \right\rangle = 2\zeta k_{\mathrm{B}}T\delta_{ij}\delta(t-t')\mathbf{1} , \tag{2.87}$$

where $\mathbf{1}$ is the unit tensor and $\boldsymbol{\xi}_i\boldsymbol{\xi}_j$ is the dyadic product[12] of $\boldsymbol{\xi}_i$ and $\boldsymbol{\xi}_j$.

[12] The dyadic product of the vectors $\boldsymbol{a}$ and $\boldsymbol{b}$ is denoted by $\boldsymbol{ab}$. This dyadic product is a second rank tensor whose ij component is defined by $(\boldsymbol{ab})_{ij} \equiv a_i b_j$, where we have put $\boldsymbol{a} = (a_i)$ and $\boldsymbol{b} = (b_i)$.

This relation guarantees that the steady state solution of (2.88) corresponds to the equilibrium canonical distribution of an ideal chain given (2.32) (see the box entitled *Fluctuation–Dissipation Theorem*). Note that (2.85) is in general nonlinear because of the flow field $\boldsymbol{v}(\boldsymbol{r}_i)$, where the variables $\{\boldsymbol{r}_i\}$ are coupled in a complex manner through the hydrodynamic interaction.

Rouse Model and its Normal Modes

If the external flow field $\boldsymbol{v}(\boldsymbol{r})$ is negligibly small, the equation of motion of a chain (2.85) reduces to

$$\zeta \frac{\partial \boldsymbol{r}_i(t)}{\partial t} = k_0 \frac{\partial^2 \boldsymbol{r}_i}{\partial i^2} + \boldsymbol{\xi}_i(t) , \tag{2.88}$$

where we have assumed that the i dependence of $\boldsymbol{r}_i$ is very small on the length scale of a segment and used the Taylor series expansion

$$\boldsymbol{r}_{i\pm 1} = \boldsymbol{r}_i \pm \frac{\partial \boldsymbol{r}_i}{\partial i} + \frac{1}{2}\frac{\partial^2 \boldsymbol{r}_i}{\partial i^2} \pm \cdots . \tag{2.89}$$

In this continuum approximation, the conditions (2.86) are rewritten as

$$\frac{\partial \boldsymbol{r}_i}{\partial i} = \boldsymbol{0} \quad \text{at } i = 0 \quad \text{and} \quad i = N , \tag{2.90}$$

which gives the boundary conditions for the partial differential equation (2.88). The dynamical model of a chain described by (2.88) and (2.90) is called the Rouse model. Because of the absence of the nonlinear coupling through the flow velocity, the equation of motion is now linear in its variable $\{\boldsymbol{r}_i(t)\}$. Due to the linearity of the Rouse model, it can easily be integrated by decomposing it into normal modes using the Fourier transformation. Such normal modes (Fourier modes) are called the Rouse modes.

For a free chain with no constraint on the chain ends, the appropriate Fourier modes are the cosine functions defined by (see Fig. 2.28a)

$$\boldsymbol{r}_i(t) = \boldsymbol{X}_0(t) + 2 \sum_{p=1,2,3,\ldots} \boldsymbol{X}_p(t) \cos \frac{p\pi i}{N} , \tag{2.91}$$

which describes a standing wave with two of its loops at the ends of the chain.[13] The inverse Fourier transformation of (2.91) is then given by

$$\boldsymbol{X}_p(t) = \frac{1}{N} \int_0^N \boldsymbol{r}_i(t) \cos \frac{p\pi i}{N} \mathrm{d}i . \tag{2.92}$$

---

[13] The normal modes for a chain one of whose ends is grafted onto a solid surface are standing waves with one node at the fixed end of the chain (see Fig. 2.28b).

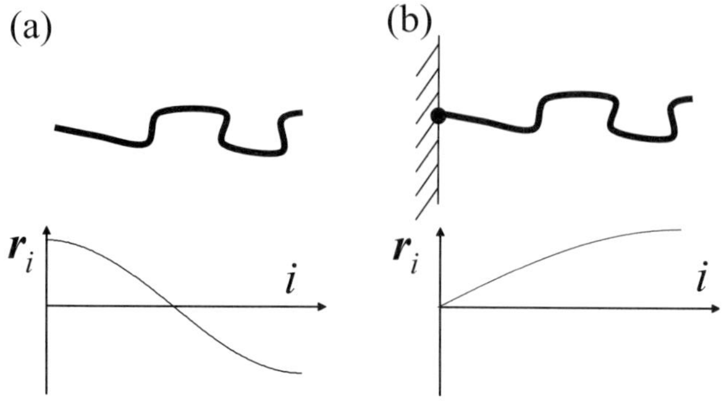

Fig. 2.28. Rouse modes with $p = 1$ for (**a**) a free chain and (**b**) a chain with one end grafted onto a solid surface

Using the Rouse modes defined in (2.91) and the boundary conditions (2.90), the equation of motion (2.88) is rewritten as

$$\frac{\mathrm{d}}{\mathrm{d}t}\boldsymbol{X}_p(t) = -\frac{k_p}{\zeta_p}\boldsymbol{X}_p(t) + \boldsymbol{\Xi}_p(t) , \tag{2.93}$$

where the parameters ζ_p and k_p are defined by

$$\zeta_p = \begin{cases} 2N\zeta & (p \neq 0) , \\ N\zeta & (p = 0) , \end{cases} \qquad k_p = \frac{2\pi^2 k_0 p^2}{N} , \tag{2.94}$$

and $\boldsymbol{\Xi}_p(t)$ is the fluctuating force acting on the pth Rouse mode defined by

$$\boldsymbol{\Xi}_p(t) = \frac{1}{\zeta N}\int_0^N \boldsymbol{\xi}_i(t)\cos\frac{p\pi i}{N}\mathrm{d}i . \tag{2.95}$$

The friction constant ζ_p has been defined so that the fluctuation–dissipation relation for the noise $\boldsymbol{\Xi}_p(t)$ has the same form as (2.87).[14] In fact, one can show using (2.87) and (2.95) that the fluctuation–dissipation relation for $\boldsymbol{\Xi}_p(t)$ is

$$\langle \boldsymbol{\Xi}_p(t)\boldsymbol{\Xi}_q(t') \rangle = \frac{2k_\mathrm{B}T}{\zeta_p}\delta_{pq}\delta(t-t')\mathbf{1} . \tag{2.96}$$

As will be shown below, the Rouse mode with $p = 0$ is a translational motion of the center of mass of the chain and is not related to the deformation of the chain conformation. The deformation of the chain is thus described by the Rouse modes with $p \geq 1$. An intuitive explanation of the Rouse modes with $p = 0$ and $p = 1$ is given in Fig. 2.29.

[14] Note that the dimension of the noise differs by a factor ζ between (2.87) and (2.95).

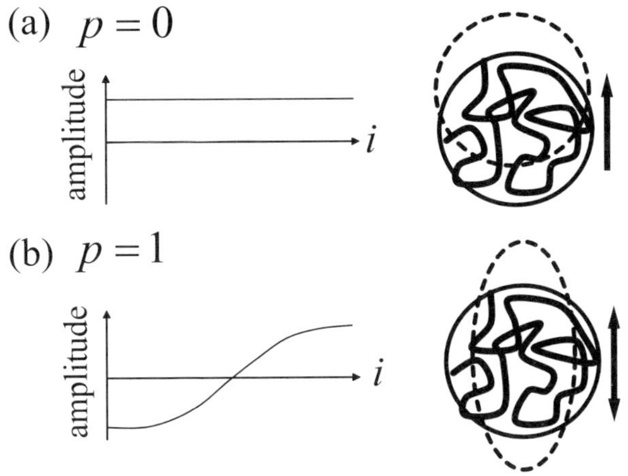

Fig. 2.29. Intuitive explanation of the Rouse modes. (**a**) The mode $p = 0$ corresponds to the translational motion of the center of mass of the chain, and (**b**) the mode $p = 1$ describes an elongation of the chain as a whole

Relaxation Phenomena in the Rouse Model

Due to the linearity of the Rouse model, the relaxation of its conformation can be described by a superposition of the relaxations of all the Rouse modes. From (2.93) and (2.94), the characteristic relaxation time of the pth Rouse mode for $p \geq 1$ is given by

$$\tau_p \equiv \frac{\zeta_p}{k_p} \propto p^{-2} . \tag{2.97}$$

On the other hand, the mode $p = 0$ is rather special. Its time evolution is given by

$$\frac{\mathrm{d}}{\mathrm{d}t}\boldsymbol{X}_0(t) = \boldsymbol{\Xi}_0(t) . \tag{2.98}$$

Using (2.92) and (2.95), we understand that the left-hand side of (2.98) is the center of mass of the chain times the chain length N, and the right-hand side is the sum of the fluctuating force acting on all the segments. Therefore, this equation shows that the center of mass of the chain undergoes Brownian motion under the influence of the thermal fluctuation. Using (2.96), (2.98) and (2.2), one can show that the diffusion constant of the center of mass of a Rouse model of length N is given by

$$D_{\mathrm{G}}^{(\mathrm{Rouse})} = \frac{k_{\mathrm{B}}T}{\zeta_0} = \frac{k_{\mathrm{B}}T}{N\zeta} \propto N^{-1} . \tag{2.99}$$

As the solvent can go freely into the coil of the Rouse chain (so-called free draining), the friction force acting on the center of mass of the chain is the sum of the friction forces acting on the individual segments. Therefore, the friction constant of the center of mass of the chain is given by $N\zeta$. Using Einstein's relation (2.84), we confirm that the diffusion constant of the center of mass of a Rouse chain is proportional to N^{-1}. In contrast to this situation, the real solvent cannot go freely into the coil of the chain due to the hydrodynamic interaction (non-draining). In such a case, the friction constant of the center of mass becomes smaller than $N\zeta$, which leads to a larger value of the diffusion constant than that predicted by the Rouse model (see Sect. 2.4.3).

We now consider the relaxation of the chain conformation. The slowest relaxation mode of the chain conformation is the Rouse mode with $p = 1$, whose characteristic relaxation time is obtained using (2.97) as

$$\tau_{\mathrm{R}} = \frac{\zeta_1}{k_1} = \frac{N^2\zeta}{\pi^2 k_0} = \frac{N^2 b^2 \zeta}{3\pi^2 k_{\mathrm{B}}T} . \tag{2.100}$$

This relaxation time is called the Rouse time.

The dependence of the viscosity η on the chain length N is also an important feature. We can evaluate it using the relation (2.103) that will be derived in the box entitled *Characteristic Behavior of Viscoelastic Bodies* on p. 71 and the Rouse relaxation time τ_{R} given in (2.100). As the Rouse chains do not interact with each other, the elasticity of a solution of Rouse chains is the sum of the contributions from the deformations of the individual chains. Thus the elasticity G is proportional to the total number of chains per unit volume, which is inversely proportional to the chain length N if the polymer volume fraction is kept constant. Therefore we have $G \sim 1/N$. Combining this result with (2.103) and (2.100), we obtain

$$\eta \sim G\tau \sim N^{-1} \times N^2 \sim N . \tag{2.101}$$

Hence, the viscosity of a solution of Rouse chains is proportional to the chain length. This chain length dependence is actually confirmed in experiments (see Fig. 5.15 in Chap. 5).

Phantom Chain Dynamics

The use of ideal chain statistics in the Rouse model leads to unrealistic dynamics in which chains can cross each other.[15] Such a model of chain dynamics is often called phantom chain dynamics (see Fig. 2.30). Although the dynamics is not very realistic, it is known that the Rouse model can reproduce the short-time-scale dynamics of polymer solutions.

Figure 2.31 shows a comparison between the theoretical prediction and the experimental results on the relaxation process of the segment density fluctuations in a polymer melt by means of the neutron spin echo method

[15] If we include excluded volume effects, the equation of motion becomes a nonlinear equation, which cannot be solved using the Fourier modes.

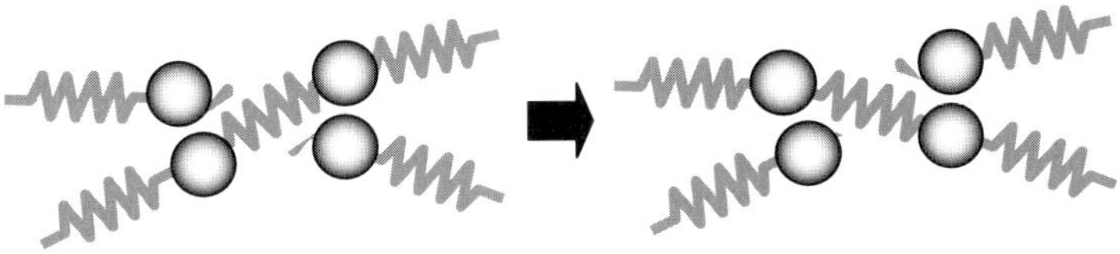

Fig. 2.30. Typical motion of phantom chains. As steric repulsion is neglected in the phantom chain model, two chains can pass through each other without entangling

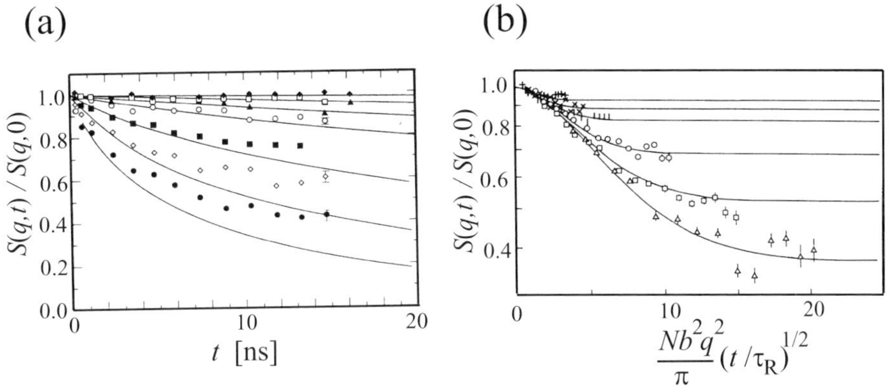

Fig. 2.31. Experimental data for the relaxation of density fluctuations in a polyisoprene melt by means of the neutron spin-echo method. (**a**) Comparison with the predictions of the Rouse model, and (**b**) comparison with those of the reptation theory by Ronca [18]. Different curves correspond to different values of the wave vector $\boldsymbol{q}$. Reprinted with permission from D. Richter et al. in *Slow Dynamics in Condensed Matter*, ed. by K. Kawasaki et al. (AIP Publishing, 1992)

(one of the inelastic neutron scattering experiment techniques). The curves indicate the intermediate scattering function, which describes the decay of the density fluctuations with wave vector $\boldsymbol{q}$. Its value at $t = 0$ corresponds to the Fourier transform of the two-body correlation function defined by (2.38) (see the box entitled *Space–Time Correlation Function* on p. 73).

Characteristic Behavior of Viscoelastic Bodies

Although we will discuss the hydrodynamic properties of viscoelastic bodies in detail in Chap. 5, we provide here a qualitative preview of their characteristic features.

A viscoelastic body has characteristics of both viscous fluids and elastic bodies. The difference between viscous fluids and elastic bodies can be observed when we continuously deform them with a constant deformation rate (amount of deformation per unit time). A viscous fluid generates a friction force against a non-zero deformation rate. Such behavior is called viscosity, and the ratio η of the friction force to the imposed deformation rate is called the coefficient of viscosity. On the other hand, an elastic body generates a restoring force against a static deformation, even if the deformation rate vanishes. This behavior is called elasticity, and the ratio G of the restoring force to the imposed deformation is called the elastic modulus. A viscoelastic body behaves as an elastic body for a (high frequency) rapid deformation, while it behaves as a viscous fluid for a (low frequency) slow deformation. Such viscoelastic behavior is measured by imposing a steady oscillatory deformation on the material and is characterized by the relation

$$\sigma(\omega) = \left[G'(\omega) + \mathrm{i}G''(\omega)\right]\kappa(\omega) , \tag{2.102}$$

where $\kappa(\omega)$ is the amplitude of the imposed oscillatory deformation with constant angular frequency ω, and $\sigma(\omega)$ is the complex amplitude of the generated force. The real part $G'(\omega)$ and the imaginary part $G''(\omega)$ are called the storage modulus and the loss modulus, respectively. The generated force associated with G' is in phase with the external deformation, while that associated with G'' has a phase delay of $\pi/2$ relative to the external deformation. For these reason, G' and G'' are related to the elasticity G and the viscosity η, respectively (see Chap. 5).

Dimensional analysis gives the characteristic time scale for such a viscoelastic relaxation τ in terms of the viscosity η and the elasticity G as

$$\tau = \eta/G . \tag{2.103}$$

The decay of the intermediate scattering function is caused by the change in the positions of segments due to diffusive motion. If the segments do not move, the intermediate scattering function will not decay. The continuous curves in Fig. 2.31a show the intermediate scattering functions calculated using the Rouse model, whilst the circles show data obtained by a neutron spin echo experiment. We can confirm that the Rouse model correctly reproduces the initial relaxation of the chain. However, the prediction of the Rouse model deviates from the experimental data as time goes on. This is because the Rouse model totally neglects the entanglements between chains, although they are crucially important in the long-time relaxation behavior of dense polymer solutions or polymer melts. In order to consider the entanglement effects correctly, we have to turn to the reptation theory, discussed in detail in Chap. 5. The theoretical predictions concerning chain relaxation based on the reptation theory are given in Fig. 2.31b.

The agreement between the experimental data and the predictions of the reptation theory is quite good, thus proving the importance of entanglement in dense polymer systems. In the reptation theory, the many-body problem of entangled chains is approximated by a single Rouse chain moving in a tube formed by the surrounding chains. Therefore, the Rouse model plays an important role in the reptation theory in spite of the fact that the Rouse model itself is not applicable to dense polymer systems.

Space–Time Correlation Function

The space correlation functions defined by (2.38) and (2.41) in Sect. 2.3.2 give a measure of the correlations between density fluctuations at different points $\boldsymbol{r}$ and $\boldsymbol{r}'$, and can be measured by neutron and X-ray scattering experiments.

In order to discuss the dynamic relaxation associated with the birth and death processes of density fluctuations in the equilibrium state, it is important to measure the following space–time correlation functions:

$$\Phi(\boldsymbol{r}-\boldsymbol{r}',t) = \frac{\mathcal{V}}{N}\sum_i\sum_j\langle\delta(\boldsymbol{r}-\boldsymbol{r}_i(t_0))\delta(\boldsymbol{r}'-\boldsymbol{r}_j(t_0+t))\rangle\,, \tag{2.104}$$

where we have assumed that the system has translational invariance in both space and time. The space–time correlation function is used when we study how the density fluctuation at time t_0 at position $\boldsymbol{r}$ causes a fluctuation at time t_0+t at another point $\boldsymbol{r}'$. In viscous fluids, as the particles can move freely by diffusion, the correlation between fluctuations decays to zero as time goes by. In elastic bodies, where average particle positions are fixed, the correlation remains non-zero even in the long-time limit $t \to \infty$.

The Fourier transform of the space–time correlation function (2.104) with respect to the space coordinates

$$S(\boldsymbol{q},t) \equiv \int \mathrm{d}\boldsymbol{r}\Phi(\boldsymbol{r},t)\mathrm{e}^{-\mathrm{i}\boldsymbol{q}\cdot\boldsymbol{r}} \tag{2.105}$$

is called the intermediate scattering function. The function $S(\boldsymbol{q},t)$ describes the relaxation behavior of the density fluctuations with wave vector $\boldsymbol{q}$. In order to measure such a time correlation function, energy is exchanged between the incident neutron (X-ray) beam and the specimen. For this purpose, inelastic scattering experiments are used.

2.4.3 Hydrodynamic Effects in Dilute Polymer Solutions

In the last section, we confirmed that the Rouse model cannot reproduce the long-time relaxation dynamics of dense polymer systems. But in that case, can the Rouse model describe chain motion in a dilute polymer solution where entanglement effects are not important? Unfortunately, the answer is negative. In dilute polymer solutions, there exists a long-range interaction between segments mediated by the flow of the solvent, which is called the hydrodynamic interaction. This long-range hydrodynamic interaction violates the basic assumption of the Rouse model, leading to non-linear dynamics that cannot be described by the Rouse model.

As shown in Fig. 2.32, when a segment moves in a solvent, it draws the surrounding solvent, and induces a local flow. Such a local flow propagates in the solvent and causes an induced flow field at a distant point, which imposes a force on the segment at the point. As a result, a long-range force is produced between the two segments. Such a long-range dynamical correlation mediated by the solvent flow is described by the so-called Oseen tensor. In this section, we derive an explicit expression for the Oseen tensor, and learn the importance of the hydrodynamic interaction for chain dynamics in dilute polymer solutions.

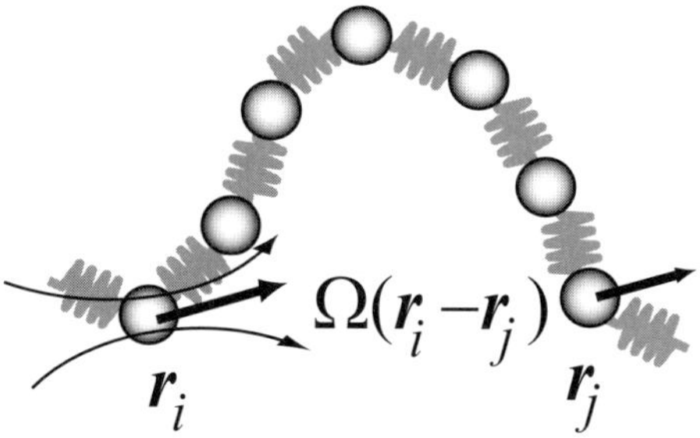

Fig. 2.32. Illustrative explanation of the hydrodynamic effects in polymer chain dynamics. When the i th segment moves, a flow is induced around it. The induced flow imposes a force on the j th segment at a distant position. Such a long-range dynamical correlation due to the hydrodynamic effect is described by the Oseen tensor

Hydrodynamic Effects and the Oseen Tensor

In order to introduce the hydrodynamic interaction into the Rouse model, we have to combine the general equation of motion (2.85) of the chain and the hydrodynamic equation of motion. Let us consider the hydrodynamics of the solvent. We denote the flow velocity of the solvent at time t and position $\boldsymbol{r}$ by $\boldsymbol{v}(\boldsymbol{r}, t)$. We impose the following three assumptions on the solvent:

- The incompressibility condition, where the density of the fluid is assumed to be constant everywhere:

$$\nabla \cdot \boldsymbol{v}(\boldsymbol{r}, t) = 0 \; . \tag{2.106}$$

- The Stokes approximation, in which we assume that the viscosity is large enough to neglect the inertia of the fluid.
- The assumption of linear viscosity, whereby the shear stress of the fluid is proportional to the gradient of the flow velocity $\nabla \boldsymbol{v}(\boldsymbol{r}, t)$.

Under these assumptions, the time evolution of the flow velocity field $\boldsymbol{v}(\boldsymbol{r}, t)$ is described by the Stokes equation

$$\frac{\partial}{\partial t}\boldsymbol{v}(\boldsymbol{r}, t) = -\frac{1}{\rho}\nabla P(\boldsymbol{r}, t) + \frac{\eta}{\rho}\nabla^2 \boldsymbol{v}(\boldsymbol{r}, t) + \boldsymbol{f}(\boldsymbol{r}, t) , \tag{2.107}$$

where ρ is the density of the fluid (constant), $P(\boldsymbol{r}, t)$ is the pressure distribution at time t, and η is the viscosity of this fluid which is assumed to be constant. The last term on the right-hand side, $\boldsymbol{f}(\boldsymbol{r}, t)$, is the external force acting on the fluid per unit mass, known as the volume force. In the polymer solution case, this volume force corresponds to the pressure imposed by the polymer segments.

Derivation of the Oseen Tensor in Incompressible Fluids

Fourier transforming the incompressibility condition (2.106) and the equation of motion (2.107), we obtain

$$\boldsymbol{q}\cdot\tilde{\boldsymbol{v}}(\boldsymbol{q}, t) = 0 \tag{2.108}$$

and

$$\frac{\partial}{\partial t}\tilde{\boldsymbol{v}}(\boldsymbol{q}, t) = -\frac{1}{\rho}\mathrm{i}\boldsymbol{q}\tilde{P}(\boldsymbol{q}, t) - \frac{\eta}{\rho}q^2\tilde{\boldsymbol{v}}(\boldsymbol{q}, t) + \tilde{\boldsymbol{f}}(\boldsymbol{q}, t) , \tag{2.109}$$

where $\boldsymbol{q}$ is the wave vector and the tilde indicates Fourier-transformed quantities. Taking the scalar product of $\boldsymbol{q}$ with both sides of (2.109) and using (2.108), we obtain

$$\tilde{P}(\boldsymbol{q}, t) = \frac{\rho\boldsymbol{q}}{\mathrm{i}q^2}\cdot\tilde{\boldsymbol{f}}(\boldsymbol{q}, t) . \tag{2.110}$$

Substituting this result into (2.109) and using the adiabatic approximation (see text) leads to

$$\frac{\partial}{\partial t}\tilde{\boldsymbol{v}}(\boldsymbol{q}, t) \approx \boldsymbol{0} = -\frac{\eta}{\rho}q^2\tilde{\boldsymbol{v}}(\boldsymbol{q}, t) + \left(\boldsymbol{1} - \frac{\boldsymbol{q}\boldsymbol{q}}{q^2}\right)\cdot\tilde{\boldsymbol{f}}(\boldsymbol{q}, t) . \tag{2.111}$$

Equation (2.114) is obtained by inverse Fourier transforming (2.111) and defining the Oseen tensor by

$$\boldsymbol{\Omega}(\boldsymbol{r}) = \frac{1}{(2\pi)^3}\int \mathrm{d}\boldsymbol{q}\frac{1}{\eta q^2}(\boldsymbol{1} - \hat{\boldsymbol{q}}\hat{\boldsymbol{q}})\mathrm{e}^{\mathrm{i}\boldsymbol{q}\cdot\boldsymbol{r}} . \tag{2.112}$$

As this inversion transformation is rather tedious, we only outline the procedure. Take the z-axis in the direction parallel to $\boldsymbol{r}$. Then the integration in (2.112) can be done using the polar coordinate to give

$$\boldsymbol{\Omega}^{(0)}(\boldsymbol{r}) = \frac{1}{8\pi\eta r}\begin{pmatrix} 1 & 0 & 0 \\ 0 & 1 & 0 \\ 0 & 0 & 2 \end{pmatrix}, \tag{2.113}$$

where the superscript (0) stands for the expression in the particular coordinate system used. Rotating the coordinate system by an angle θ around the y-axis and then by an angle φ around the z-axis, the above tensor is transformed into the desired form given by (2.115). (Note that $\boldsymbol{\Omega}$ obeys the transformation rule for second rank tensors.)

If the relaxation of the solvent flow is much faster than the relaxation of the chain conformation, we can assume that the flow field of the solvent is always in equilibrium under the external force field $\boldsymbol{f}(\boldsymbol{r})$ imposed by the polymer segments. Under this assumption, we can solve (2.107) to obtain the solvent flow velocity. Such an approximation is called the adiabatic approximation. As (2.107) is linear, this calculation is easily performed to give

$$\boldsymbol{v}(\boldsymbol{r}, t) = \int \mathrm{d}\boldsymbol{r}' \boldsymbol{\Omega}(\boldsymbol{r} - \boldsymbol{r}') \cdot \rho \boldsymbol{f}(\boldsymbol{r}', t) \,. \tag{2.114}$$

Details are given in the box entitled *Derivation of the Oseen Tensor in Incompressible Fluids*. Here, $\boldsymbol{\Omega}$ is called the Oseen tensor defined by

$$\boldsymbol{\Omega}(\boldsymbol{r}) \equiv \frac{1}{8\pi\eta r}\left(\mathbf{1} + \frac{\boldsymbol{r}\boldsymbol{r}}{r^2}\right) . \tag{2.115}$$

The Oseen tensor shows that the thermodynamic force acting on the fluid induces a flow that is inversely proportional to the distance from the point. As explained in the box entitled *Projection Operator and Oseen Tensor*, the object $\mathbf{1} + \boldsymbol{r}\boldsymbol{r}/r^2$ on the right-hand side of (2.115) is the projection operator which eliminates the longitudinal component of the flow velocity $\boldsymbol{v}(\boldsymbol{r})$. This is a consequence of the incompressibility condition of the solvent which prohibits density waves such as sound waves.

Projection Operator and Oseen Tensor

Given a constant unit vector $\hat{\boldsymbol{a}}$ ($|\hat{\boldsymbol{a}}| = 1$), one can decompose any vector $\boldsymbol{u}$ into its components parallel and perpendicular to $\hat{\boldsymbol{a}}$, denoted $\boldsymbol{u}_{\parallel}$ and $\boldsymbol{u}_{\perp}$, respectively:

$$\boldsymbol{u}_{\parallel} = (\boldsymbol{u} \cdot \hat{\boldsymbol{a}})\hat{\boldsymbol{a}} \,,$$

$$\boldsymbol{u}_{\perp} = \boldsymbol{u} - \boldsymbol{u}_{\parallel} = (\mathbf{1} - \hat{\boldsymbol{a}}\hat{\boldsymbol{a}})\cdot\boldsymbol{u} \,.$$

The second equation shows that the operator (tensor) $\mathbf{1} - \hat{\boldsymbol{a}}\hat{\boldsymbol{a}}$ eliminates the component parallel to $\hat{\boldsymbol{a}}$ from any vector. Such an operator is called a projection operator.

Equation (2.111) shows that the Oseen tensor eliminates the longitudinal component of the flow velocity parallel to its propagating direction, i.e., the velocity component parallel to $\boldsymbol{q}$, which is inhibited by the incompressibility condition.

Equation of Motion of a Chain with Hydrodynamic Interaction

If we denote the total free energy of the polymer solution by F, the j th segment of the polymer chain imposes the thermodynamic force $\boldsymbol{F}_j = \partial F/\partial \boldsymbol{r}_j$ on the solvent. As this thermodynamic force acts on the solvent only at position $\boldsymbol{r}_j$, the volume force $\rho\boldsymbol{f}(\boldsymbol{r},t)$ in (2.114) can be written as

$$\rho\boldsymbol{f}(\boldsymbol{r},t) = \sum_j \boldsymbol{F}_j \delta(\boldsymbol{r} - \boldsymbol{r}_j) \,. \tag{2.116}$$

Combining this equation with (2.114), we obtain the expression

$$\boldsymbol{v}(\boldsymbol{r},t) = \sum_j \boldsymbol{\Omega}(\boldsymbol{r} - \boldsymbol{r}_j)\cdot\boldsymbol{F}_j(t) \tag{2.117}$$

for the flow field. Then, substituting this expression into $\boldsymbol{v}(\boldsymbol{r},t)$ in (2.85), we obtain

$$\frac{\partial}{\partial t}\boldsymbol{r}_i(t) = \frac{1}{\zeta}\boldsymbol{\mu}_i(t) + \sum_{j(\neq i)} \boldsymbol{\Omega}(\boldsymbol{r}_i - \boldsymbol{r}_j)\cdot\boldsymbol{F}_j(t) + \frac{1}{\zeta}\boldsymbol{\xi}_i(t) \,, \tag{2.118}$$

where $\boldsymbol{\mu}_i = k_0\big(\boldsymbol{r}_{i-1} + \boldsymbol{r}_{i+1} - 2\boldsymbol{r}_i\big)$ is the force due to the two bonds connected to the i th segment, defined by $\partial\mathcal{H}_0/\partial\boldsymbol{r}_i$. In the summation in the second term on the right-hand side, we excluded the contribution from $i = j$ because this term accounts for the self-interaction. The fluctuation–dissipation relation for the fluctuating force is given by

$$\langle \boldsymbol{\xi}_i(t)\boldsymbol{\xi}_j(t')\rangle = \begin{cases} 2\zeta k_{\mathrm{B}} T \mathbf{1}\delta(t-t') & (i = j) \,, \\ \boldsymbol{\Omega}(\boldsymbol{r}_i - \boldsymbol{r}_j)\delta(t-t') & (i \neq j) \,, \end{cases} \tag{2.119}$$

which is obtained by applying the fluctuation–dissipation relation to (2.118). The above model equation with the hydrodynamic interaction (2.118) cannot be solved analytically because of the nonlinearity in the Oseen tensor. One of the methods for solving such a nonlinear equation is to use a direct computer simulation technique. An example of such a numerical simulation is shown in Fig. 2.47 in Sect. 2.7.

Flow Field Screening in the Presence of the Hydrodynamic Interaction

The hydrodynamic interaction between the segments induces a counter flow that cancels the external flow inside the coil of the chain. Hence, the external flow can go only partially into the coil. This phenomenon is called screening of the flow field. When the external flow field is screened, the coil of the chain behaves as an impenetrable solid sphere (see Fig. 2.33). If the friction due to the external fluid acting on the chain is equal to the friction on a rigid sphere with radius R_{hydro}, this R_{hydro} is called the radius of the hydrodynamically equivalent sphere.

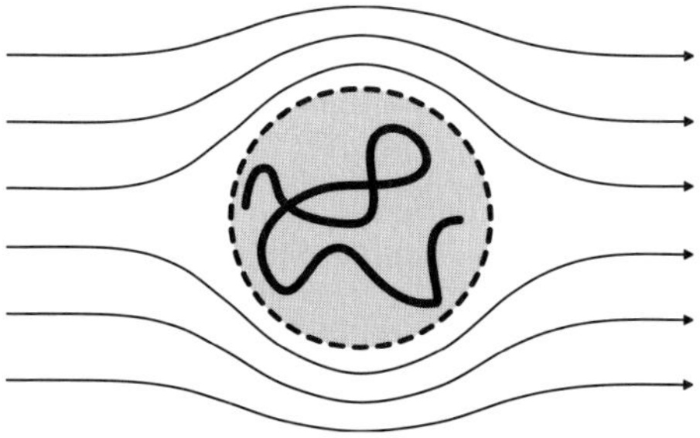

Fig. 2.33. Due to the screening effect, the external flow field cannot go into the coil of the chain

Zimm Model and Relaxation Phenomena

Next we consider an approximate analytical solution of the nonlinear equation of motion (2.118). Due to the hydrodynamic interaction, the model predicts different relaxation behavior to the Rouse model discussed in Sect. 2.4.2. As (2.118) is a nonlinear equation in which segments couple nonlinearly through the Oseen tensor, it cannot be solved using the Fourier modes. In order to linearize this equation, we replace $\boldsymbol{\Omega}(\boldsymbol{r}_i - \boldsymbol{r}_j)$ by its average $\left\langle \boldsymbol{\Omega}(\boldsymbol{r}_i - \boldsymbol{r}_j) \right\rangle_0$ over the conformational distribution of an ideal chain or an excluded volume chain. This approximation is a kind of mean field approximation, called the pre-averaging approximation. Such a chain model is called the Zimm model. Once we have averaged the Oseen tensor, the equation of motion (2.118) becomes a linear equation which can be solved using the normal modes, as was done in the case of the Rouse model. Here, we skip the details of the calculation and only show the results [6]:

$$\text{diffusion constant of center of mass} \qquad D_{\mathrm{G}} \sim \frac{k_{\mathrm{B}}T}{\eta b} N^{-\nu} ,$$

$$\text{longest relaxation time} \qquad \tau_1 \sim \frac{\eta b^3}{k_B T} N^{3\nu} , \tag{2.120}$$

where η is the viscosity of the solvent. The parameter ν is the exponent of the end-to-end distance, which takes the value $\nu = 1/2$ for an ideal chain [see (2.9)] and $\nu = 3/5$ for an excluded volume chain [see (2.56)], respectively. Compared with the results (2.99) and (2.100) for the Rouse model, the diffusion becomes faster and the relaxation time becomes shorter in the long chain limit. The hydrodynamic interaction generates dynamical correlations between distant segments and tends to align the segment velocity in the same direction. This effect accelerates the diffusion and the relaxation of the chain conformation.

The importance of the hydrodynamic interaction introduced in the Zimm model can be confirmed by measuring the storage modulus $G'(\omega)$ and loss modulus $G''(\omega)$ experimentally. Here the elastic moduli G' and G'' have been introduced in (2.102). Figure 2.34 compares the predictions of the Rouse and Zimm models for $G'(\omega)$ and $G''(\omega)$ and the corresponding experimental

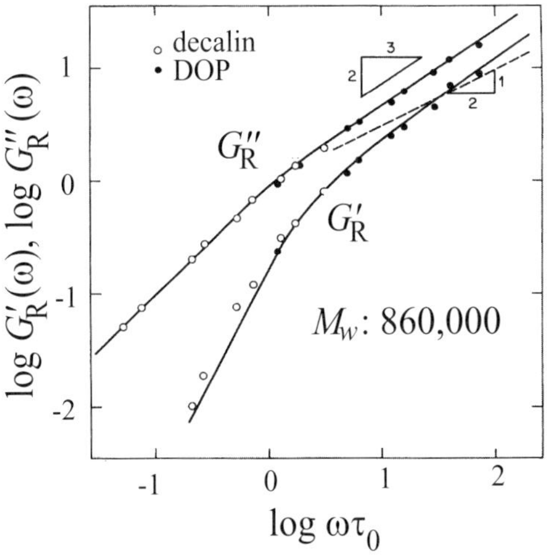

Fig. 2.34. Comparison between the theoretical prediction of the frequency-dependent elastic moduli based on the Zimm model and the corresponding experimental data for the dilute θ-solution of polystyrene. G'_R and G''_R are reduced moduli defined by $G'_R \equiv \dfrac{M}{Nk_BT} \lim_{c\to 0} \dfrac{G'}{c}$ and $G''_R \equiv \dfrac{M}{Nk_BT} \lim_{c\to 0} \dfrac{G'' - \omega\eta_s}{c}$, where M and N are the molecular weight and the number of segments per chain, respectively, η_s is the viscosity of the solvent, and c is the polymer mass density. The parameter τ_0 is a characteristic relaxation time of the chain. The predicted slope in the high frequency region is 2/3 for the Zimm model and 1/2 for the Rouse model. The experimental result supports the Zimm model. Reprinted with permission from R.M. Johnson et al., Polym. J. **1**, 742 (1970). (© 1970, Polymer Society Japan)

results with a θ-solution of polystyrene. From this figure, we can confirm that the analytic results based on the Zimm model can reproduce the response of a dilute polymer solution to an external field very well.

2.5 Justification of the Gaussian Chain Model from a Microscopic Point of View

2.5.1 Full Atomistic Model of Polymer Chains

In the preceding sections, we derived a bead–spring model by applying the coarse-graining procedure to a lattice model. We then investigated the statistical properties of polymer chains using this model. The time has come to discuss how realistic the assumptions used in the derivation of the bead–spring model are from a microscopic point of view.

Figure 2.35 illustrates the hierarchy of descriptions of polymer systems on different length and time scales ranging from microscopic to macroscopic scales. In this hierarchy, the most microscopic description is the one shown in Fig. 2.35a, where the polymer chain is observed on the atomic length scale 0.1 nm. On such a scale, one could see the individual carbon atoms and hydrogen atoms that compose the polymer chain. Such a description of the polymer chain using all the component atoms is called the full atomistic model. The dynamics of the chain on this length scale are completely described by Newton's equations of motion for the individual atoms. (In this

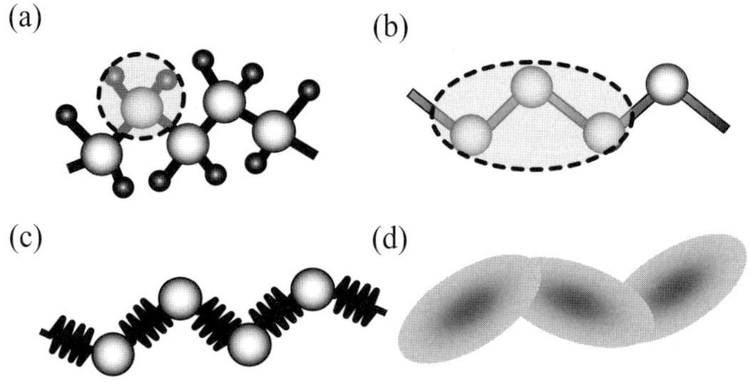

Fig. 2.35. Different descriptions of a polymer chain from the microscopic scale to longer length scales. (**a**) The scale of individual atoms ($\sim$ 0.1 nm). (**b**) The scale of atomic groups consisting of several atoms ($\sim$ 1 nm). (**c**) The scale of segments defined by a set of many atomic groups defined in (**b**), where the polymer chain is described as a sequence of segments connected by harmonic bonds (1–10 nm). (**d**) Much longer scales where individual segments are no longer distinguished and distribute continuously ($\geq$ 10 nm)

book, we do not consider quantum effects. Of course, quantum descriptions should be used for electrical, optical, and sub-atomic scale properties.)

In principle, we can predict any macroscopic phenomena by solving these coupled equations of motion. However, typical time scales of the conformational relaxation of polymer chains in dense solutions or in melts are of the order 10^{-6}–10^{-3} s, while the characteristic time scale of the motion of an individual atom shown in Fig. 2.35a is of the order 10^{-12} s. Moreover, the polymer system often exhibits much slower phenomena, such as phase separation and crystallization, whose characteristic time scales can be hours or days. Hence, the ratio between the microscopic time scales and the mesoscopic/macroscopic time scales is sometimes of the order 10^{13}–10^{15}. It is hard to cover such huge time scales by a fully atomistic model. To overcome this difficulty, we have to go beyond the full atomistic model, with regard to both the length scale and the time scale.

2.5.2 United Atom Model

When there are coexisting phenomena with widely different space–time scales, the coarse-graining method discussed in Sect. 2.2.1 is effective. One such example is the model shown in Fig. 2.35b, which is valid on the length scale of 1 nm. On this scale, the fast vibrations of the chemical bonds connecting the carbon atoms and the hydrogen atoms can be averaged out, and we can treat the CH_2 atomic group as a single rigid body. Such a model is called the united atom model. In this model, the microscopic details and the chemical properties of the monomers are introduced through the interaction potentials between the united atoms. As illustrated in Fig. 2.36, the following interactions are usually assumed:

- bond length (bond stretching),
- angle between two adjacent bonds (bond bending),
- angle between two planes formed by three consecutive bonds (torsion),
- contact interaction between pairs of united atoms (excluded volume interactions, etc.).

Among the interactions shown in Fig. 2.36, it is mainly bond bending and torsion that cause directional correlations between adjacent bonds along the chain, i.e., the consecutive –C–C–C– bonds prefer some particular conformations as a result of the anisotropy in the valence bonds and the steric hindrance between the atomic groups connected to the carbon atoms. Two typical stable conformations of the –C–C–C– bonds are the trans conformation and the gauche conformation (see Fig. 2.37).

As the trans conformation has lower potential than the gauche conformation, the trans conformation is dominant at low temperature. In such a sequence of trans conformations, all carbon atoms are placed on the same plane with a zigzag shape. As a result, the chain lies straight in a certain direction (Fig. 2.38a). On the other hand, the gauche conformation has larger

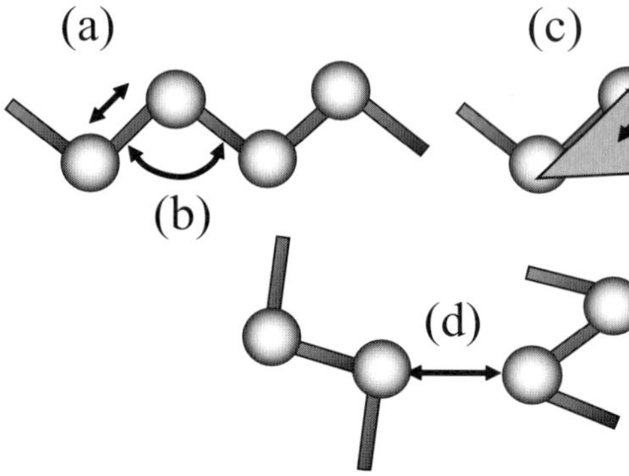

Fig. 2.36. Interactions taken into account in the united atom model of neutral polymers. (**a**) Bond stretching interaction, (**b**) bond bending interaction, (**c**) torsion interaction, and (**d**) contact interaction between segments

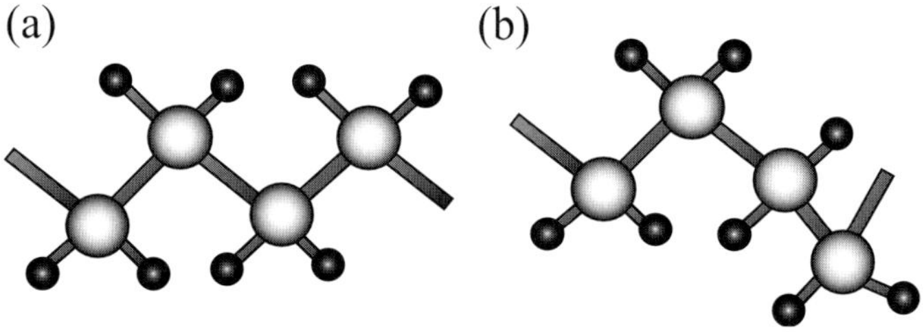

Fig. 2.37. Two stable conformations of the –C–C–C– bonds in the backbone. (**a**) The trans conformation and (**b**) the gauche conformation. *Large* and *small spheres* show carbon atoms and hydrogen atoms, respectively

potential than the trans conformation. Hence, gauche conformations appear from place to place in the sequence of trans conformations, and destroy the directional correlations of the backbone (Fig. 2.38b). The characteristic correlation length of the directional order of the backbone is called the persistence length l_{p}, beyond which the directional correlation of the backbone is lost.

Persistence Length and Kuhn Length

Instead of the persistence length, the Kuhn length b_{K} is often used as a measure of the rigidity of the chain:

$$b_{\mathrm{K}} \equiv \frac{\langle |\boldsymbol{R}|^2 \rangle}{R_{\mathrm{max}}} .$$

This has dimensions of length. Here $\boldsymbol{R}$ is the end-to-end vector and R_{max} is the maximum length to which the chain can be stretched.

Let us show that the Kuhn length and the persistence length are equivalent. We assume that the chain is composed of N segments whose size is taken as the persistence length. As the directions of the bonds are statistically independent, the square of the end-to-end distance is given by

$$\langle |\boldsymbol{R}|^2 \rangle = N l_{\mathrm{p}}^2 \, .$$

On the other hand, since the segment behaves as an almost rigid rod on the scale of the persistence length, the maximum length of the chain is

$$R_{\max} = N l_{\mathrm{p}} \, .$$

Dividing both sides of these two equations, we immediately recognize that the persistence length l_{p} is equal to the Kuhn length.

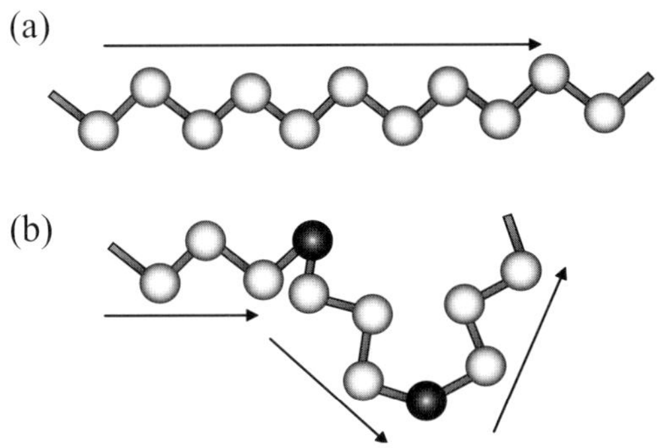

Fig. 2.38. Directional correlations between bonds along the backbone. (**a**) All-trans conformation and (**b**) a conformation containing a few gauche conformations. *Black atoms* show the positions of the gauche conformations and *arrows* indicate the direction of the order of the backbone

We further apply the coarse-graining procedure to this united atom model. Let us define a coarse-grained segment in such a way that its size is greater than the persistence length l_{p}. Then the directions of the bond vectors connecting such coarse-grained segments are statistically independent. If the system is isotropic, the probability distribution function of the coarse-grained bond vector $\boldsymbol{u}$ is given by

$$P(\boldsymbol{u}) = P(|\boldsymbol{u}|) \, , \tag{2.121}$$

which is a function of only the magnitude of $\boldsymbol{u}$. Applying the same procedure as was used to derive (2.20)–(2.26), we finally obtain the Gaussian distribution (2.26). (After all, any bond distribution functions reduce to Gaussian chain behavior on a longer length scale, as long as excluded volume interactions are absent.) This argument also shows that the persistence length l_{p} is the minimum length scale above which the chain can be regarded as a Gaussian chain. In polymer science, the term Kuhn length is often used instead of the persistence length (see the box entitled *Persistence Length and Kuhn Length*).

2.6 Statistical Theories and Experiments on Semi-Flexible Chains

Examples of Semi-Flexible Chains

As discussed in the last section, any microscopic model reduces to the Gaussian chain model characterized by (2.32), provided that the segment size is chosen to be larger than the persistence length and excluded volume effects are neglected. Such a Gaussian chain model describes the universal properties of long polymer chains. One of the important extensions of this Gaussian chain model toward microscopic scales is the worm-like chain model, in which the backbone has a finite bending elasticity.

If we define the segment size to be smaller than the persistence length, there remain directional correlations between adjacent bonds. A similar situation also arises if there is a long-range repulsive Coulomb interaction between charged segments in the chain. In these cases, the polymer chain adopts a locally straight conformation. Such a chain is said to be worm-like or semi-flexible. Typical examples of the worm-like chain are the polyelectrolyte, DNA and actin filaments.

As actin filaments and DNA molecules are especially long, i.e., their typical chain length is of the order 100 nm–1 μm, the motion of these chains can be directly observed using a fluorescence microscope (see Fig. 1.6). For this reason, DNA chains and actin filaments are useful targets for experimental tests of dynamical theories of polymer chains. On the other hand, the chain conformations of DNA chains and actin filaments are not fully described by Gaussian chain statistics because of the bending elasticity. In this section, we extend the statistical theory for the flexible Gaussian chains discussed so far to worm-like chains, and give a brief review of experimental tests of the theory of worm-like chains using DNA and actin filaments.

2.6.1 Worm-Like Chain Model

In order to describe the statistical mechanics of worm-like chains (WLC), we add an extra term to the Hamiltonian of the Gaussian chain model:

$$\mathcal{H}_{\mathrm{WLC}} = \mathcal{H}_0 + \mathcal{W}_{\mathrm{bend}} , \tag{2.122}$$

where $\mathcal{H}_0$ is the Hamiltonian of a flexible ideal Gaussian chain and $\mathcal{W}_{\mathrm{bend}}$ is the bending elastic energy. So far, we have been using the Gaussian chain model, i.e., the first term on the right-hand side of (2.30), for the ideal chain Hamiltonian $\mathcal{H}_0$. A more suitable choice of $\mathcal{H}_0$ for worm-like chains is the Hamiltonian of a freely-jointed chain, which is a chain of segments joined by freely-rotating rigid bonds. As will be described in Exercise 2.2, the freely-jointed chain model exhibits ideal Gaussian chain behavior on coarse-grained scales.

On the basis of this freely-jointed chain model, one can construct a simple model for the bending elastic energy $\mathcal{W}_{\text{bend}}$ by assuming a harmonic potential for the angle between two adjacent bond vectors. Then $\mathcal{W}_{\text{bend}}$ is written in terms of the bond vectors $\{\boldsymbol{u}_i\}$ as

$$\mathcal{W}_{\text{bend}} = \frac{\kappa_0}{2}\sum_{i=0}^{N-1} |\hat{\boldsymbol{u}}_i - \hat{\boldsymbol{u}}_{i+1}|^2 = -\kappa_0 \sum_{i=0}^{N-1} \hat{\boldsymbol{u}}_i \cdot \hat{\boldsymbol{u}}_{i+1} + \text{const.} \,, \tag{2.123}$$

where κ_0 is the bending elastic modulus, $\hat{\boldsymbol{u}}_i \equiv \boldsymbol{u}_i/|\boldsymbol{u}_i|$ is the unit vector parallel to the i th bond vector, and the constant is independent of the chain conformation. In the continuum limit where the discrete index i is replaced by a continuous variable s, the Hamiltonian for the bending elastic energy (2.123) leads to

$$\mathcal{H}_{\text{WLC}} \to \frac{\kappa}{2}\int_0^L \left|\frac{\mathrm{d}\hat{\boldsymbol{t}}}{\mathrm{d}s}\right|^2 \mathrm{d}s = \frac{\kappa}{2}\int_0^L c(s)^2 \mathrm{d}s \,, \tag{2.124}$$

where we have introduced the scaled bending elastic modulus $\kappa = b\kappa_0$ and the contour length of the chain $L = Nb$. Furthermore, $\boldsymbol{t}(s) \equiv \mathrm{d}\boldsymbol{r}(s)/\mathrm{d}s$ is the tangent vector to the chain, $\hat{\boldsymbol{t}}$ is the unit vector parallel to $\boldsymbol{t}$, and $c(s)$ is the curvature of the contour of the chain defined by

$$c(s) \equiv \left|\frac{\mathrm{d}\hat{\boldsymbol{t}}(s)}{\mathrm{d}s}\right| \qquad \left[= \left|\frac{\mathrm{d}\theta(s)}{\mathrm{d}s}\right| \quad \text{in the 2-dimensional case}\right] . \tag{2.125}$$

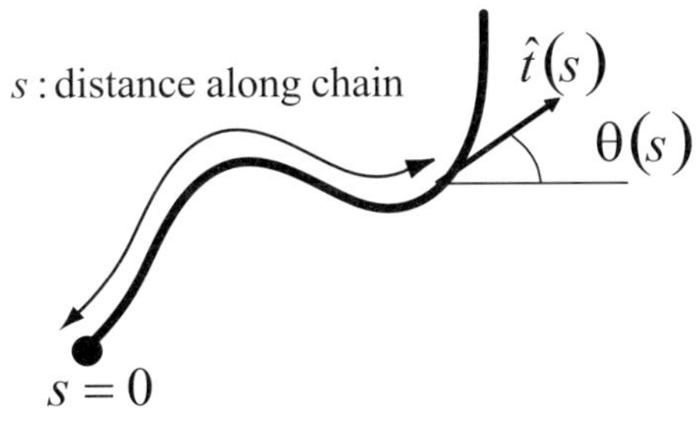

Fig. 2.39. Schematic picture of a two-dimensional worm-like chain. The chain is locally straight due to the bending elasticity and hence follows a smooth curve

Curvature of Two- and Three-Dimensional Curves

Consider a curve $\boldsymbol{r}(s)$, where s is a parameter specifying a position on the curve. The unit tangent vector of this curve at s is given by

$$\hat{\boldsymbol{t}}(s) = \frac{\mathrm{d}\boldsymbol{r}(s)}{\mathrm{d}s} \bigg/ \left| \frac{\mathrm{d}\boldsymbol{r}(s)}{\mathrm{d}s} \right| .$$

If we define s as the distance along the chain measured from a specified point on the curve (such a coordinate s is called the natural coordinate), the denominator of the above equation simplifies to $|\mathrm{d}\boldsymbol{r}(s)/\mathrm{d}s| = 1$. Then the derivative of $\hat{\boldsymbol{t}}(s)$ with respect to s

$$\boldsymbol{n}(s) \equiv \frac{\mathrm{d}\hat{\boldsymbol{t}}(s)}{\mathrm{d}s}$$

gives the rate of change of the unit tangent vector and is a normal vector to the curve. The curvature is defined as the magnitude of $\boldsymbol{n}(s)$, i.e.,

$$c(s) \equiv \left| \frac{\mathrm{d}\hat{\boldsymbol{t}}(s)}{\mathrm{d}s} \right| \quad \left[= \left| \frac{\mathrm{d}\theta(s)}{\mathrm{d}s} \right| \quad \text{in the 2-dimensional case} \right] .$$

In deriving the last equation, we used the fact that the unit vector for a two-dimensional curve is given by $\big(\cos\theta(s), \sin\theta(s)\big)$, where $\theta(s)$ is the angle between the tangent vector and the positive direction of the x-axis.

In deriving the above expression for the curvature, we used the fact that the index i can be identified with the distance along the chain (see the box entitled *Curvature of Two- and Three-Dimensional Curves*). The last line on the right-hand side of (2.125) is an expression for a two-dimensional chain, where $\theta(s)$ is the angle between the unit tangent vector $\hat{\boldsymbol{t}}(s)$ and the direction of an arbitrarily selected x-axis (see Fig. 2.39).

When the local curvature is small, the model (2.123) is approximated by the bending elasticity, proportional to the square of the curvature, as shown in (2.124).

2.6.2 Statistical Properties of a Stretched Worm-Like Chain

Although a worm-like chain tends to have a locally straight conformation, it behaves in the same manner as a flexible chain on length scales longer than the persistence length. The chain thus adopts a globular random-coil conformation unless any external force is present. When we stretch the chain by imposing a force on both ends, a restoring tension is generated. Such tension can be measured for a DNA chain using the laser-tweezers technique, which can be a crucial test for statistical theories of worm-likc chains (see the box entitled *Laser-Tweezers Technique*).

Laser-Tweezers Technique

The laser-tweezers technique is a technique to fix a particle at a desired position or to force it to migrate using a laser beam. The mechanism of the laser-tweezers technique is explained schematically in Fig. 2.40. Consider a particle in a solvent whose refractive index is smaller than that of the solvent. When a narrow laser beam is emitted towards this particle, the laser beam is refracted at the surface of the particle, and this imposes a force on the particle in reaction to the momentum change caused by the refraction. Integrating this force over the whole surface of the particle leads to a restoring force on the particle towards the central axis of the laser beam, and this is used to fix the particle position. By changing the position of the beam, one can force the particle to migrate in the solvent. The advantage of using a laser beam is that it does not perturb the solvent flow. Since the contrast between the diffractive indices of the water and the DNA molecule is not large enough, a small particle made of polystyrene is attached to one end of the DNA to improve the contrast.

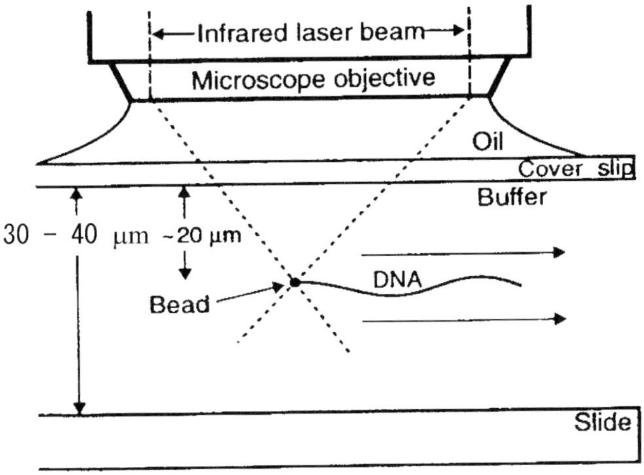

Fig. 2.40. Schematic illustration of the laser-tweezers method. A DNA molecule is dissolved in a solvent that fills the space between two parallel glass plates. A laser beam is then focused at an appropriate position between the plates where the attractive potential is generated. Reprinted with permission from T.T. Perkins et al., Science **264**, 822 (1994). (© 1994, AAAS)

We first consider a theoretical treatment of the worm-like chain described by the Hamiltonian (2.124) [19]. We set the x-axis along the direction of chain stretching. Denoting the tension acting on both ends of the chain by F, the Hamiltonian of this stretched chain is given by

$$\mathcal{H} = \frac{\kappa}{2}\int_0^L \left|\frac{\mathrm{d}\hat{\boldsymbol{t}}(s)}{\mathrm{d}s}\right|^2 \mathrm{d}s - F\hat{\boldsymbol{x}}\cdot\int_0^L \hat{\boldsymbol{t}}(s)\mathrm{d}s \,, \tag{2.126}$$

where $\hat{\boldsymbol{x}}$ is the unit vector along the x-axis and $\hat{\boldsymbol{t}}(s)$ is the unit tangent vector at s. Here we note that $\hat{\boldsymbol{t}}(s) = \mathrm{d}\boldsymbol{r}(s)/\mathrm{d}s$ because we use the natural coordinate s (see the box entitled *Curvature of Two- and Three-Dimensional Curves*).

When the chain is highly stretched in the x-direction, $\hat{\boldsymbol{t}}$ is almost parallel to the x-axis. We write $\hat{\boldsymbol{t}} = (t_{\parallel}, \boldsymbol{t}_{\perp})$, where $t_{\parallel}$ and $\boldsymbol{t}_{\perp}$ are the components of $\hat{\boldsymbol{t}}$ parallel and perpendicular to the x-axis, respectively. Then, using $|\hat{\boldsymbol{t}}| = 1$ and expanding $\mathcal{H}$ as a power series in $\boldsymbol{t}_{\perp}$, we obtain

$$\begin{aligned} \hat{\boldsymbol{x}}\cdot\hat{\boldsymbol{t}} \equiv t_{\parallel} &= \sqrt{1-|\boldsymbol{t}_{\perp}|^2} \approx 1 - \frac{1}{2}|\boldsymbol{t}_{\perp}|^2 \,, \\ \left|\frac{\mathrm{d}\boldsymbol{t}}{\mathrm{d}s}\right|^2 &= \left(\frac{\mathrm{d}t_{\parallel}}{\mathrm{d}s}\right)^2 + \left|\frac{\mathrm{d}\boldsymbol{t}_{\perp}}{\mathrm{d}s}\right|^2 \approx \left|\frac{\mathrm{d}\boldsymbol{t}_{\perp}}{\mathrm{d}s}\right|^2 \,, \end{aligned} \tag{2.127}$$

up to second order in $\boldsymbol{t}_{\perp}$. Substituting (2.127) into (2.126), we obtain the harmonic Hamiltonian

$$\begin{aligned} \mathcal{H} &\approx \int_0^L \left(\frac{\kappa}{2}\left|\frac{\mathrm{d}\boldsymbol{t}_{\perp}}{\mathrm{d}s}\right|^2 + \frac{F}{2}|\boldsymbol{t}_{\perp}|^2\right)\mathrm{d}s - FL \\ &= 2L\sum_q \frac{1}{2}\left(\kappa q^2 + F\right)|\boldsymbol{a}_q|^2 - FL \,, \end{aligned} \tag{2.128}$$

where $\boldsymbol{a}_q$ is the Fourier transformation of $\boldsymbol{t}_{\perp}$ defined similarly to (2.92):

$$\boldsymbol{a}_q = \frac{1}{L}\int_0^L \boldsymbol{t}_{\perp}\cos(qs)\,\mathrm{d}s \,. \tag{2.129}$$

The inverse transformation to (2.129) is given by

$$\boldsymbol{t}_{\perp}(s) = 2\sum_q \boldsymbol{a}_q \cos(qs) \,, \tag{2.130}$$

where $\boldsymbol{a}_0 = \boldsymbol{0}$ holds due to the fact that the end-to-end displacement in the yz direction vanishes.

As we approximated the Hamiltonian up to the second order in the variables in (2.128), the canonical distribution of the variables is Gaussian and the average over such a distribution is easily obtained by performing Gaussian integrals. If we assume that the chain is in equilibrium at temperature T, the equipartition theorem requires

$$2L \times \frac{1}{2}(\kappa q^2 + F)\langle |\boldsymbol{a}_q|^2 \rangle = d \times \frac{k_{\mathrm{B}}T}{2} , \tag{2.131}$$

where the factor d on the right-hand side is the dimensionality. Using this relation, we obtain

$$\frac{1}{L}\int_0^L \langle |\boldsymbol{t}_\perp(s)|^2 \rangle \mathrm{d}s = 2\sum_q \langle |\boldsymbol{a}_q|^2 \rangle = \frac{1}{\pi}\int_0^\infty \mathrm{d}q \frac{k_{\mathrm{B}}Td}{\kappa q^2 + F} = \frac{k_{\mathrm{B}}Td}{2\sqrt{\kappa F}} . \tag{2.132}$$

Integrating the first equation of (2.127) with respect to s over the interval $0 \le s \le L$, we can obtain the extension of the chain in the x-direction. Then, performing the integration in (2.132), we obtain the following relation between the chain extension x and the tension F:

$$\frac{x}{L} \approx 1 - \frac{1}{2L}\int_0^L \mathrm{d}s \langle |\boldsymbol{t}_\perp(s)|^2 \rangle = 1 - \sqrt{\frac{k_{\mathrm{B}}T}{4l_{\mathrm{p}}F}} , \tag{2.133}$$

where we have used the relation $\kappa = k_{\mathrm{B}}Tl_{\mathrm{p}}$ [see (2.136) in the box entitled *Relation between Bending Elasticity Modulus and Persistence Length*], and we have assumed $d = 2$, considering the microscopy images as the target. Solving (2.133) for F, we obtain

$$F = \frac{k_{\mathrm{B}}T}{l_{\mathrm{p}}} \frac{1}{4(1 - x/L)^2} . \tag{2.134}$$

As we consider a highly stretched chain that is almost parallel to the x-axis, the expression (2.134) is valid for a large deformation of a worm-like chain. On the other hand, if the deformation is small, the chain should behave as an ideal chain with potential (2.28). Interpolating the result in the large deformation limit (2.134) and that in the small deformation limit (2.28), we obtain the interpolation formula

$$\frac{l_{\mathrm{p}}}{k_{\mathrm{B}}T}F = \frac{x}{L} + \frac{1}{4(1 - x/L)^2} - \frac{1}{4} . \tag{2.135}$$

In the flexible Gaussian limit, the persistence length l_{p} is related to the effective bond length b through the relation (2.140). Using this relation, we can confirm that (2.135) reduces to (2.28) for $x/L \ll 1$. As will be described in the next section, this formula for the chain extension gives a good description of the properties of biopolymers such as DNA chains.

Relation Between Bending Modulus and Persistence Length

The bending modulus κ defined in (2.123) and the persistence length l_{p} introduced in Sect. 2.5 are related to each other through

$$\kappa = b\kappa_0 = k_{\mathrm{B}}Tl_{\mathrm{p}} . \tag{2.136}$$

Let us prove this relation.

Using the Hamiltonian for the bending elasticity (2.123), the equilibrium probability distribution of the angle between two adjacent bond vectors $\hat{\boldsymbol{u}}_i$ and $\hat{\boldsymbol{u}}_{i+1}$ is given by

$$\langle \hat{\boldsymbol{u}}_i \cdot \hat{\boldsymbol{u}}_j \rangle = \langle \cos\theta \rangle = \frac{\int \mathrm{d}\Omega \cos\theta \exp(\beta\kappa_0 \cos\theta)}{\int \mathrm{d}\Omega \exp(\beta\kappa_0 \cos\theta)} = 1 - \frac{1}{\beta\kappa_0} , \tag{2.137}$$

where $\int \mathrm{d}\Omega$ is the integration over the steric angle. The definition of the persistence length means that the directional correlation between bond orientations is lost beyond l_{p}/b segments along the chain, which leads to the relation

$$\langle \hat{\boldsymbol{u}}_i \cdot \hat{\boldsymbol{u}}_j \rangle = \exp\left(-\frac{|i-j|b}{l_{\mathrm{p}}}\right) . \tag{2.138}$$

Substituting $j = i+1$ in this expression, we obtain

$$\langle \hat{\boldsymbol{u}}_i \cdot \hat{\boldsymbol{u}}_{i+1} \rangle = \exp\left(-\frac{b}{l_{\mathrm{p}}}\right) \sim 1 - \frac{b}{l_{\mathrm{p}}} , \tag{2.139}$$

where we have assumed $b \ll l_{\mathrm{p}}$. Comparing (2.139) with (2.137), we immediately obtain the relation (2.136).

When the chain is flexible enough, the directions of individual bonds become statistically independent. In such a case, the persistence length l_{p} is essentially the same as the effective bond length b and we obtain the relation (see Exercise 2.3)

$$2l_{\mathrm{p}} = b \qquad \text{(flexible chain)} . \tag{2.140}$$

2.6.3 Experiments on Worm-Like Chains Using Biopolymers

DNA and actin filaments are typical biopolymers. These molecules are so large that one can observe them directly with optical microscopes. In fact, using the fluorescent microscopy technique, the motion of individual chains has been extensively studied. The most remarkable difference between actin filaments and DNA molecules is the ratio between their chain length and their persistence length. Typical physical properties concerning the bending elasticity of actin filaments and DNA molecules are summarized in Table 2.1. In actin filaments, the persistence length is in general of the same order as the chain length. Thus an actin filament behaves like a semi-flexible rod. On the other hand, the persistence length of a DNA molecule is much smaller than its chain length, and it adopts a random coil structure in water.

Table 2.1. Typical physical properties concerning the bending elasticity of conventional polymers, DNAs and actin filaments

Chain type	Chain length L [nm]	Persistence length l_p [nm]	L/l_p
Conventional polymer	10^2–10^3	~ 1	$\geq 10^2$
DNA	10^4–10^5	~ 50	10^2–10^3
Actin filament	$\sim 10^4$	$\sim 10^3$	10^1

Measurement of the Bending Modulus of Actin Filaments

A series of fluorescence microscope images of conformational changes in an actin filament are shown in Fig. 2.41. It is obvious from these pictures that the actin filament is a rigid molecule and can be expected to be a good candidate for application of the formula (2.124). We consider an experiment on the bending properties of an actin filament, where the validity of the model Hamiltonian (2.124) is examined.

As the Hamiltonian (2.124) is in the quadratic form of the chain conformation, it can be rewritten as a superposition of the contributions from

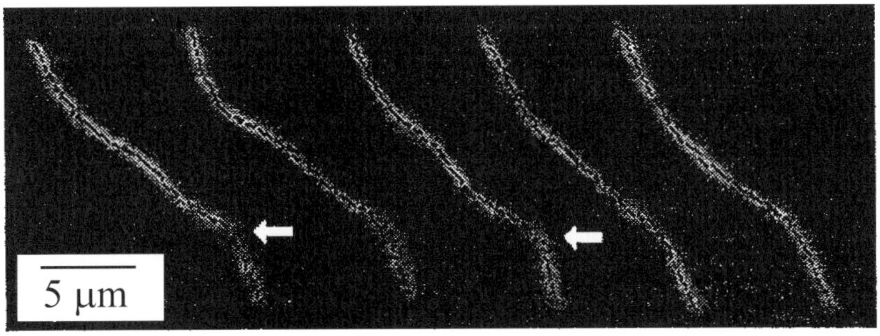

Fig. 2.41. Fluorescence microscope images of temporal conformational changes in an actin filament. Reprinted with permission from J. Käs et al., Europhys. Lett. **21**, 865 (1993). (© 1993, EDP Sciences)

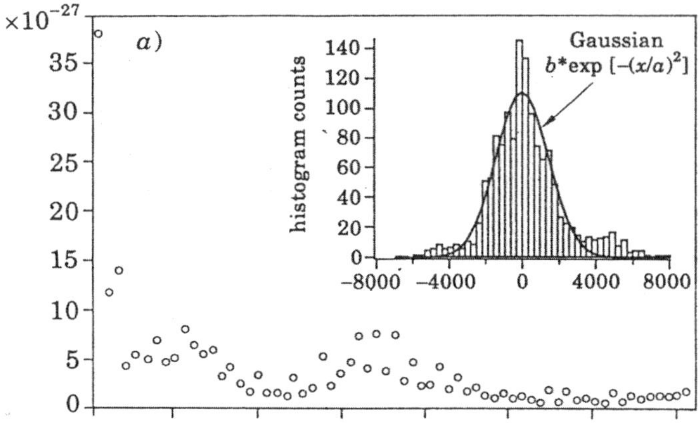

Fig. 2.42. Wave number dependence of the bending elastic modulus κ of an actin filament determined from the Fourier spectrum of its conformations as obtained by the fluorescence microscopy technique. The bending elastic modulus κ is estimated as $\kappa = (4.0 \pm 0.4) \times 10^{-27}$ J m. Reprinted with permission from J. Käs et al., Europhys. Lett. **21**, 865 (1993). (© 1993, EDP Sciences)

independent Fourier modes. This is done in a similar manner to (2.128) to give

$$\mathcal{H}_{\mathrm{WLC}} = \kappa L \sum_{q>0} q^2 a_q^2 \,, \tag{2.141}$$

where L is the chain length and a_q is the Fourier transformation of $\theta(s)$ defined in Fig. 2.39 by

$$\theta(s) = \theta_0 + \sum_{q>0} a_q \cos(qs) \,. \tag{2.142}$$

Assuming that each Fourier mode a_q gains energy $k_{\mathrm{B}}T/2$ according to the equipartition theorem [similarly to (2.131)], we obtain the relation

$$\langle a_q^2 \rangle = \frac{dk_{\mathrm{B}}T}{2\kappa L q^2} \,. \tag{2.143}$$

As the left-hand side of the relation can be obtained from the numerical Fourier transform of the fluorescence microscopy images, we can evaluate the bending modulus κ experimentally. In Fig. 2.42, we show the bending elasticity modulus κ obtained using the above method as a function of the wave number q. For a wide range of wave numbers, κ is constant, irrespective of the wave number, which justifies the worm-like chain treatment for actin filaments.

Behavior of a DNA Molecule in a Flow Field

Next, we discuss experiments on the elastic and the relaxation properties of the conformation of DNA molecules using the laser-tweezers technique.

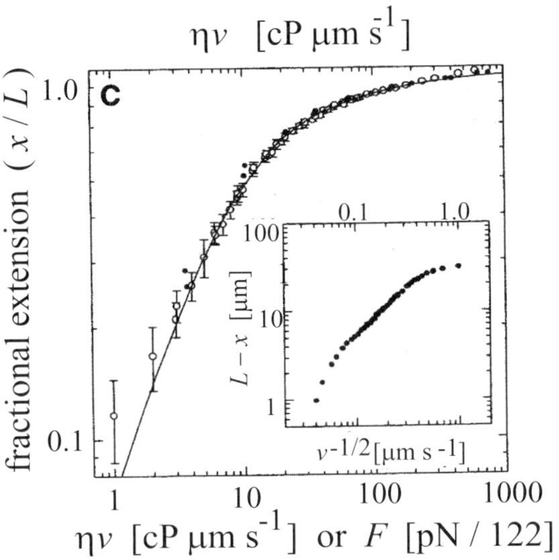

Fig. 2.43. Relation between the flow velocity v and the scaled extension (x/L) of a DNA chain in a uniform flow field. *Symbols* show the experimental data while the *continuous curve* shows the theoretical curve given by (2.135) with the assumption $F \propto v$. For a highly stretched chain, (2.135) predicts $v \propto F \propto (L-x)^{-2}$, which leads to $L - x \sim v^{-1/2}$. The *inset* shows the relation between $L - x$ and $v^{-1/2}$, which shows that the second term on the right-hand side of (2.135) deviates from the experimental behavior in the large extension region. Reprinted with permission from T.T. Perkins et al., Science **268**, 83 (1995). (© 1995, AAAS)

We fix one end of a DNA molecule using the laser-tweezers technique and impose a uniform flow to stretch the DNA molecule (Fig. 1.6a). We then obtain the extension of the chain denoted by x as a function of the chain length L and the flow velocity v. As the chain is highly stretched, the screening effect due to the hydrodynamic interaction is expected to be weak. The friction force F acting on the segment is thus approximately proportional to the flow velocity v, i.e., $F \propto v$. Figure 2.43 shows the relationship between the flow velocity and the extension of the DNA chain. In this figure, the scaled extension x/L, i.e., the chain extension x divided by the chain length L, is plotted as a function of the product of the viscosity of the solvent η and the flow velocity v. The continuous curve shows the result of the theory using the worm-like chain model given in (2.135). We find that the worm-like chain model adequately reproduces the stretching behavior of the DNA chain.[16]

[16] The agreement between theory and experiment does not simply mean that the worm-like chain model is valid for the DNA molecule in a flow field. This is because of the different situations between them. The theoretical formula (2.135) is derived by assuming that both ends of the chain are pulled by an extensional

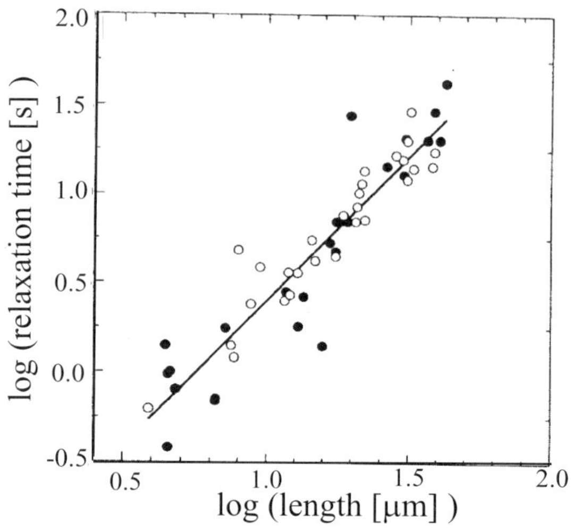

Fig. 2.44. The longest relaxation time of the conformation of an elongated DNA chain as a function of the chain length. Reprinted with permission from T.T. Perkins et al., Science **264**, 822 (1994). (© 1994, AAAS)

We now focus on the relaxation dynamics of the conformation of a DNA molecule in a solvent. As explained in Sect. 2.4.3, the dynamics of such a DNA molecule is expected to be described by the Zimm model, which includes the effects of the solvent flow. Figure 2.44 shows the results of an experiment on the relaxation of an elongated DNA molecule one of whose end points is fixed by the laser tweezers and which is subjected to a uniform flow. When the flow is switched off, the chain starts to shrink. Figure 2.44 shows the double logarithmic plot of the longest relaxation time of the chain conformation as a function of the chain length.

This result shows that the longest relaxation time τ_1 and the chain length N are related via $\tau_1 \sim N^{1.66}$, which is close to the predicted behavior of the Zimm model ($\tau_1 \sim N^{1.8}$) for an excluded volume chain with Flory exponent $\nu = 3/5$ [see (2.120)]. (Because DNA is a polyelectrolyte, its local conformation is nearly straight due to the electrostatic repulsion between segments with the same charge. Therefore, the segment of a DNA chain is not a spherical body, as is assumed in the Zimm model, but an elongated rod-like object. In such a case, the excluded volume effect becomes smaller than in the case with spherical segments, and the exponent ν for the extension of the coil becomes smaller than the Flory exponent $\nu = 3/5$ of a typical excluded volume

force while the experimental data is obtained by imposing an extensional force on each segment via the external flow.

chain. Due to the difference in segment shape, the hydrodynamic interaction is also modified.)

2.7 Molecular Simulations of Polymer Dynamics

2.7.1 Molecular Simulation Methods

In this chapter, we have been studying statistical theories of chain conformation and its dynamics, which were checked by experiments using biopolymers. Another useful way of verifying the theories on complex polymer systems is computer simulation. Analytic theories are often useless for systems with many interacting chains or systems with highly deformed chains. In such cases, computer simulations can give useful information. In this section, we discuss several examples of computer simulations based on molecular models of polymer chains.

The typical molecular models of polymer chains used in computer simulations are extensions of the stochastic bead–spring model introduced in Sect. 2.4.2 or extensions of the bead–spring model with hydrodynamic flow effects introduced in Sect. 2.4.3. To make the model more realistic, the excluded volume interaction and the non-linearity in the spring potential are often introduced. The molecular dynamics method (MD), the Monte Carlo method (MC), and the Brownian dynamics method (BD) are used to simulate these mesoscopic molecular models [20,21]. As each of these methods can generate a series of samples of chain conformations from the equilibrium or non-equilibrium statistical ensembles, physical quantities can be calculated from these samples.

In the MD method, the chain dynamics is simulated by numerically integrating Newton's equations of motion for all segments or atoms. In the BD method, instead of Newton's equations of motion, we numerically integrate the Langevin equation (2.88) which includes the thermal fluctuation effect. Both the MD and BD simulations can simulate the time evolution of the system as well as the equilibrium state. On the other hand, the MC method generates a sequence of samples from the canonical ensemble (or other ensembles). In general, the sample sequence of the MC method does not correspond to the time evolution of the system. Thus, the MC method basically simulates static equilibrium properties.[17]

[17] The MC method is also used for dynamics or for non-equilibrium states. In such cases, the updating procedure of the MC method is itself assumed to define the dynamics of the system. As long as the MC procedure is appropriately chosen so that the MC trial move is limited only to local conformational changes, the time evolution of the system on the coarse-grained level does not contradict the real dynamics.

2.7.2 Models of Interaction Potentials for Coarse-Grained Chains

In performing molecular simulations such as MD, MC or BD simulations, an inappropriate choice of the interaction potential between segments may lead to an incorrect result, not only for the microscopic behavior but also for the mesoscopic universal behavior. Although the bead–spring model described by (2.88) is useful in analytical treatments, it has several shortcomings when used in molecular simulations. For example, (2.88) does not include the following properties:

- the excluded volume interaction,
- the entanglement between chains,
- the finite extensibility of the bonds.

The absence of the excluded volume effect results in an unrealistic compressibility of polymer melts, and the lack of the entanglement effect totally alters the long-time dynamics of dense polymer systems.

In molecular simulations, the excluded volume effect is introduced by assuming the LJ potential defined by (2.3) between segment pairs. In order to guarantee the entanglement effect, we have to replace the harmonic spring by the following nonlinear spring with finite extensibility [21]:

$$U_{\mathrm{bond}}(r) = \begin{cases} -\dfrac{1}{2}kr_{\mathrm{max}}^2 \ln\left[1-\left(\dfrac{r}{r_{\mathrm{max}}}\right)^2\right] & (r < r_{\mathrm{max}})\,, \\ \infty & (r > r_{\mathrm{max}})\,. \end{cases} \tag{2.144}$$

This nonlinear spring potential is called the Finite Extensible Nonlinear Elastic (FENE) potential. The FENE potential reduces to the usual harmonic potential in the small extension region, while it diverges at the distance r_{max} causing the finite extensibility. The entanglement effect can be realized by choosing an appropriate value for the particle diameter σ compared with the maximum extension r_{max}. Thus the bead–spring model with the LJ potential and the FENE potential is often used as a coarse-grained model of both dilute and dense polymer systems. This model overcomes all the shortcomings listed above.

2.7.3 Examples of Molecular Simulations

An example of the BD simulation introduced in the previous section is shown in Fig. 2.45. In this simulation, 20 chains each composed of 200 segments are contained in a cubic box, and the Langevin equation for the individual segment is integrated numerically to trace the time evolution of the whole system.[18] These figures show the temporal evolution of a single tagged chain

[18] In order to eliminate the finite size effects of the simulation box, the so-called periodic boundary condition is used, where the two parallel faces of the simulation box are assumed to be identical so that the particle going out of one face is coming into the system through the other face.

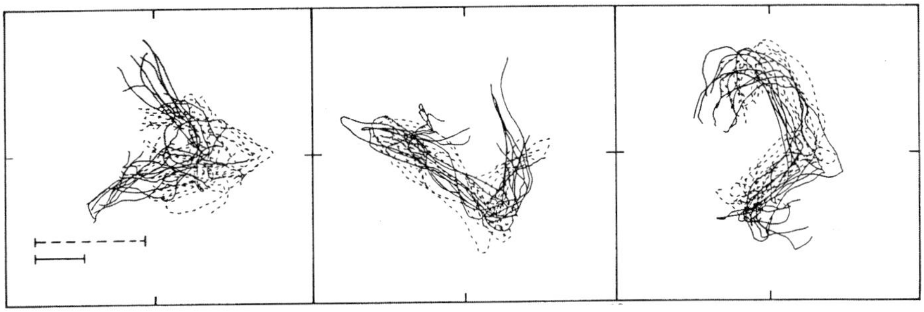

Fig. 2.45. Examples of the temporal change in the chain conformation obtained by an BD simulation on a dense polymer melt using the bead–spring model. The system is a cubic box containing 20 chains, each of which is composed of 200 segments. Reprinted with permission from K. Kremer and G.S. Grest, J. Chem. Phys. **92**, 5057 (1990). (© 1990, American Institute of Physics)

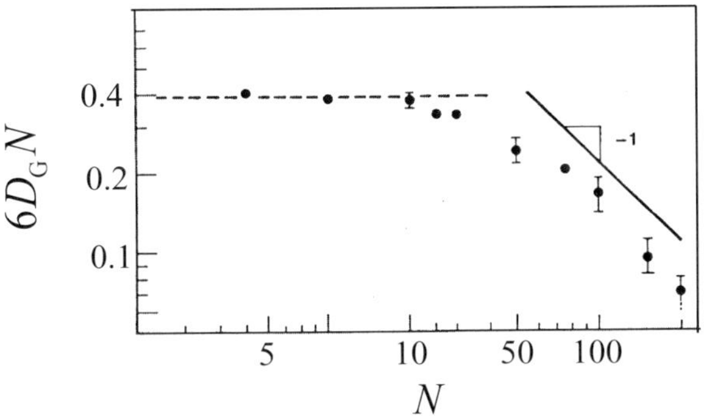

Fig. 2.46. Relation between the center-of-mass diffusion constant D_G and the chain length N (number of segments per chain) obtained from the BD simulation shown in Fig. 2.45. Reprinted with permission from K. Kremer and G.S. Grest, J. Chem. Phys. **92**, 5057 (1990). (©1990, American Institute of Physics)

among the 20 chains. As the tagged chain cannot move freely due to entanglements with the other chains, it mainly moves along its chain contour. Such motion is called reptation.

Figure 2.46 shows the simulation results for the dependence of the diffusion constant of the center of mass of the chain D_G on the chain length N (number of segments per chain). For short chains with $N < 50$, the Rouse behavior ($D_G \sim N^{-1}$) given by (2.99) is observed. When the chain becomes longer, the behavior changes to $D_G \sim N^{-2}$. The behavior of the diffusion coefficient found in the MD simulation coincides with the predictions of the reptation theory, to be discussed in Chap. 5.

By adding the hydrodynamic effects described by (2.118) to the bead–spring model with the FENE potential and the LJ potential, it is possible

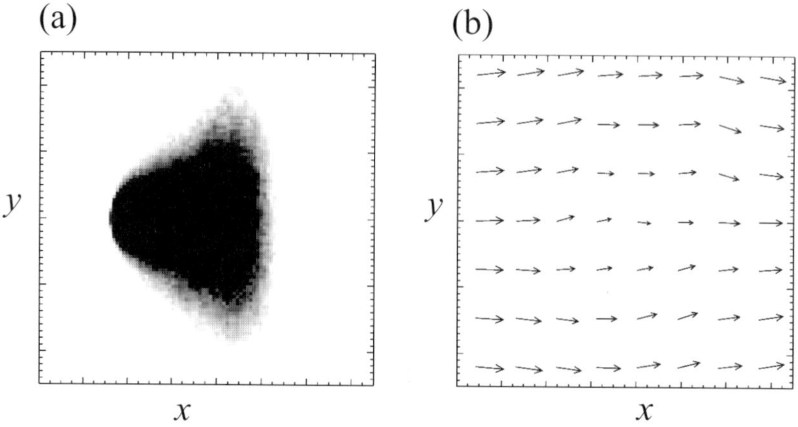

Fig. 2.47. A BD simulation of a polymer chain with one end fixed and subjected to a uniform external flow. The hydrodynamic interaction is taken into account using the Oseen tensor (see Fig. 2.43 for the experimental result). (**a**) The segment distribution and (**b**) the distribution of the flow velocity of the solvent. One can confirm that the external flow field is excluded from the coil due to the hydrodynamic screening effect. Reprinted with permission from R. Rzehak et al., Europhys. Lett. **46**, 821 (1999). (© 1999, EDP Sciences)

to simulate a chain in a solvent under the influence of the hydrodynamic interaction. An example of such simulations is shown in Fig. 2.47, which corresponds to the elongated DNA chain in a uniform flow shown in Fig. 1.6a. One can check that the segment distribution is deformed due to the external flow field. Here, the important effect of the hydrodynamic interaction is the screening effect, which excludes the external flow from the inside of the coil. Such a screening effect is also observed in this simulation.

If we want to simulate microscopic properties of a polymer system, a more microscopic chain model such as the united atom model introduced in Sect. 2.5 is used with realistic interatomic potentials determined by quantum mechanical calculations. Such a model is useful for studying the microscopic dynamics and statistics of polymer chains, such as crystallization, tacticity effects, etc.

In conclusion, the molecular simulation of polymers is a powerful tool for studying both static equilibrium properties and dynamic properties of polymer systems, taking the chain architecture and chemical details into account. However, to perform simulations, we have to integrate a large set of coupled equations of motion for the individual segments or atoms. This limits the use of molecular simulations to long-time behavior or large-scale systems. In order to eliminate such a difficulty, we have to rely on further coarse-grained models, to be discussed in Chaps. 3–5.

Exercises

Exercise 2.1. Show that the gyration radius of an ideal chain composed of $N+1$ segments is given by (2.10). Hint: First show that

$$\sum_{i=0}^{N} |\boldsymbol{r}_i - \boldsymbol{r}_\mathrm{G}|^2 = \frac{1}{2(N+1)} \sum_{i=0}^{N} \sum_{j=0}^{N} |\boldsymbol{r}_i - \boldsymbol{r}_j|^2 \qquad (2.145)$$

and use the property given in (2.8).

Exercise 2.2. As shown in Fig. 2.48, consider a chain model composed of segments connected by freely rotating rigid rods with length b_0. If we neglect excluded volume effects, show that the probability distribution of the end-to-end vector of this chain reduces to (2.26) in the long-chain limit. Hint: Assume that the distribution of each bond vector is isotropic. Then, instead of the distribution function (2.12) for the lattice model, the bond distribution function is given by

$$P_1(\boldsymbol{u}) = \frac{1}{4\pi b_0^2} \delta\big(|\boldsymbol{u}| - b_0\big) \,, \qquad (2.146)$$

where 4π is the total steric angle.

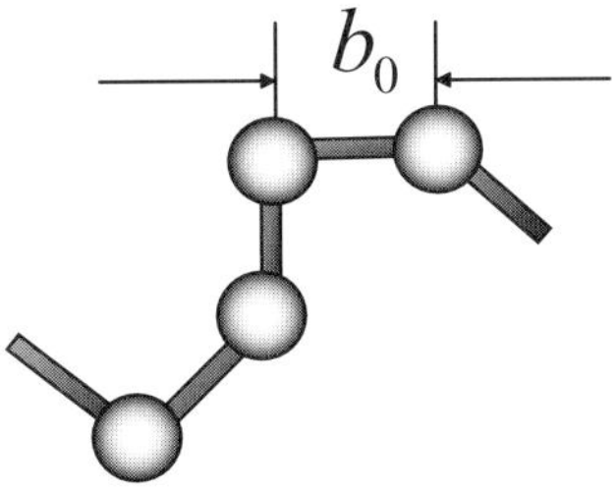

Fig. 2.48. Freely-jointed chain model

Exercise 2.3. For a flexible chain, let us demonstrate that the relation (2.140) holds between the persistence length l_p and the effective bond length b. First of all, use (2.138) to derive the relation

$$\big\langle |\boldsymbol{R}|^2 \big\rangle \sim 2 l_\mathrm{p} b N \qquad (N \gg 1) \,. \qquad (2.147)$$

Comparing this relation with $\big\langle |\boldsymbol{R}|^2 \big\rangle \sim N b^2$, one can obtain (2.140).

3 Mesoscopic Structures and Self-Consistent Field Theory

The previous chapter was devoted to exploring molecular models of polymers. In order to construct a suitable model for large-scale phenomena, we now introduce a further coarse-grained model based on the self-consistent field theory. The self-consistent field theory is a technique to calculate segment distributions and probability distributions of chain conformations within the mean field approximation. This technique is especially useful for quantitative estimation of dense inhomogeneous polymer systems, such as the phase separation structures of polymer mixtures and the chain conformations near an interface. It is extensively used to design and predict the physical properties of polymer materials and polymer composites.

3.1 Mesoscopic Phenomena in Polymer Systems

One of the essential factors that determine the characteristic size of the phenomena in polymer systems is the gyration radius of the constituent polymer chains. Although the size of a monomer is of the order of 1 nm, the gyration radius of a highly polymerized chain can reach 10–10^2 nm. The structures and dynamics on the length scale of the gyration radius are therefore clearly separated from those on the atomic or monomer length scale.

A typical example of such phenomena is the phase separation of polymer blends and block copolymers. A mixture of two kinds of highly polymerized chains is usually immiscible because of the extremely low mixing entropy of the polymer chains compared to the system composed of low-molecular-weight particles. This tendency is clearly expressed in the Flory–Huggins free energy (2.67). While the entropy S contains a factor $1/N_K$ ($K = A$ or B), such a factor does not appear in the interaction energy E. The factor $1/N_K$ originates from the fact that the mixing of two types of polymer is driven by the translational degrees of freedom of the center of mass of the individual chain, which is only three per chain irrespective of the chain length, while the segment interaction per chain is proportional to the chain length N_K. Thus, even a very small chemical difference between the monomers is exaggerated in polymer systems, and this leads to a phase separation.

Figure 3.1 illustrates typical phase-separated domain structures in a mixture of homopolymers (polymer blend) (Fig. 3.1a) and a melt of diblock

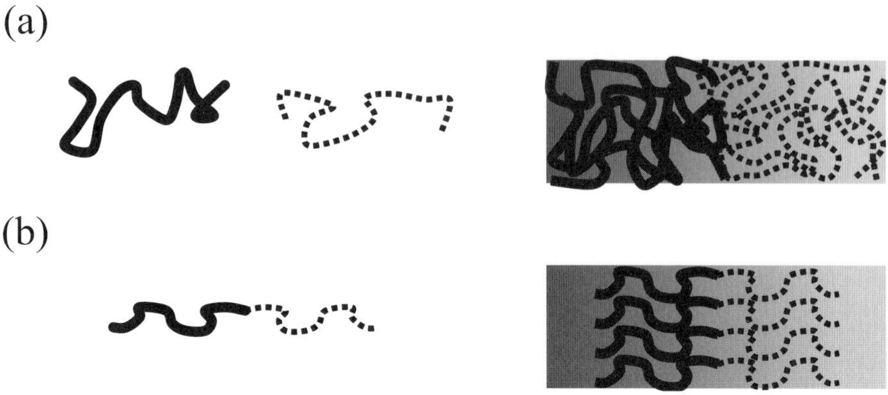

Fig. 3.1. Phase separation phenomena in polymer systems. (**a**) An immiscible polymer blend that shows macrophase separation. (**b**) A block copolymer melt that shows a microphase separation

copolymer composed of two types of homopolymer connected by chemical bonds (Fig. 3.1b). In the case of the block copolymer melt, a macroscopic phase separation cannot take place because the two polymer chains are chemically bonded. Instead, smaller domains are generated on the same scale as the gyration radius. (This is the so-called weak segregation regime, where the domain boundary is not sharp. In the strong segregation regime, where the boundary of the domain is sharp, the domain size can be much larger than the gyration radius of the chain. We shall discuss such a case in Sect. 3.3.4.) Such a phase separation is called a microphase separation. The tendency towards phase separation is specified by the parameter χN, i.e., the χ-parameter defined in (2.72) times the chain length N. It can be proved that the status of the system depends on χ and N only through the parameter χN for polymer melts. This can be done by transforming the basic equation to a dimensionless form (see Exercise 3.1).

As a typical example of the phase separations of polymers, Fig. 3.2 shows electron microscope images of the domain structure of a microphase-separating polystyrene–polyisoprene (PS–PI) block copolymer melt.

Figure 3.3 shows the experimentally observed phase diagram of the microphase separation of PS–PI block copolymer melts, which is the same block copolymer as in Fig. 3.2. The melt is uniformly mixing in the high temperature region (small χN), while various microphase separations are observed in the low temperature region (large χN). [Note that the vertical axis χN is inversely proportional to the temperature T due to the definition of the χ-parameter in (2.68) and (2.72).] The solid circle indicates the critical point, where the difference between the PS-rich phase and the PI-rich phase disappears.

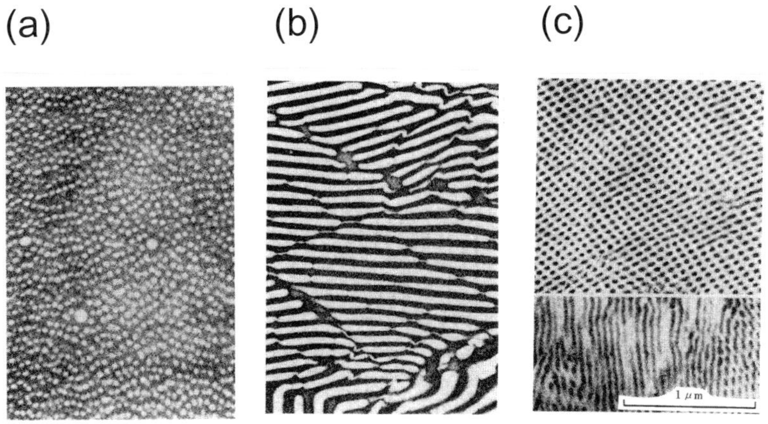

Fig. 3.2. Transmission electron microscope images of the microphase separation of a polystyrene–polyisoprene (PS–PI) block copolymer melt. The block ratio PS/PI and the observed domain structure are (**a**) PS/PI = 20/80 (spherical domains of PS immersed in a matrix of PI), (**b**) PS/PI = 40/60 (lamellar domains composed of alternating PS and PI layers), and (**c**) PS/PI = 60/40 (cylindrical domains of PI in a matrix of PS). Reprinted with permission from Koubunshi Sozai Shashinshuu (in Japanese) (© 1992, Polymer Society of Japan)

Figure 3.4 gives an example of molecular simulation of the microphase separation phenomena. The figure shows an equilibrium state obtained by a Brownian dynamics (BD) simulation of a cubic system that contains 200 AB-type diblock copolymer chains, each of which is composed of $50A$ segments and $50B$ segments. (For details of the molecular simulations, see Sect. 2.7.) The interaction parameter between the AB segment pair is chosen approximately as $\chi = 1.0$. Figure 3.4a clearly shows a layered microphase-separated domain structure (a lamellar structure).

Figure 3.4b also shows the coarse-grained segment density distributions in the perpendicular direction to the layers. On a larger length scale, the essential properties of the lamellar structure are retained in these coarse-grained segment density distributions. The self-consistent field theory is based on such a coarse-grained description using the segment density distribution.

3.2 Formulation and Simple Examples of the Self-Consistent Field Theory of Polymers

In this section, we introduce a powerful theoretical method for calculating the phase separation structures and interfacial structures of dense polymer systems. It is called the self-consistent field (SCF) theory. In the SCF theory, the mean field approximation is used to reduce the many-chain problem to a

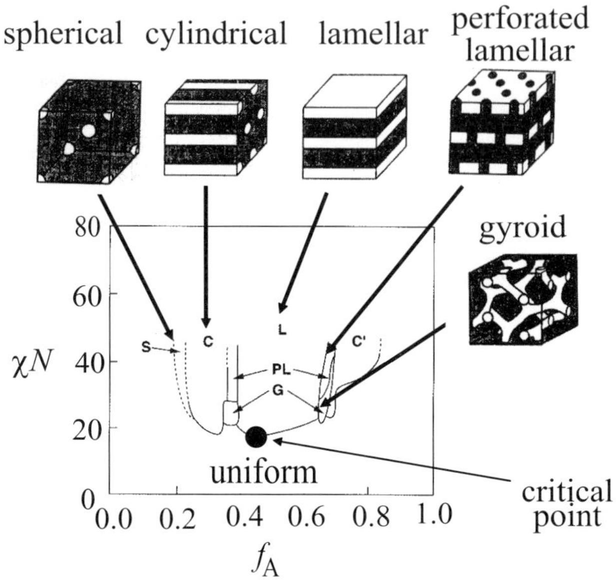

Fig. 3.3. The phase diagram of the microphase separation of polystyrene–polyisoprene (PS–PI) block copolymer melt. The *ordinate* is the product χN of the interaction parameter and the chain length, and the *abscissa* is the so-called block ratio, i.e., the ratio of the length of the PS block to the total length of the block copolymer chain. Reprinted with permission from F.S. Bates and G.H. Fredrickson, Physics Today **52**, 32 (1999). (© 1999, American Institute of Physics)

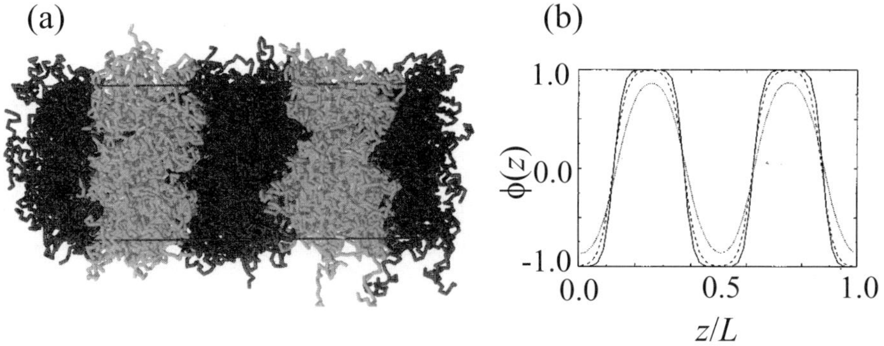

Fig. 3.4. A lamellar microphase separation structure obtained by a Brownian dynamics simulation of an AB-type diblock copolymer melt. (**a**) Snapshot of the chain conformations ($\chi = 1.0$), and (**b**) segment density distributions averaged over a direction parallel to the layers: $\chi = 1.0$ (*continuous curve*), 0.5 (*broken curve*), and 0.3 (*dotted curve*). Reprinted with permission from M. Murat, G.S. Grest and K. Kremer, Macromolecules **32**, 595 (1999). (© 1999, American Chemical Society)

single-chain problem in an external field, where the chain conformation of the tagged chain is calculated using the path integral method. We first formulate the SCF method, and then give several simple examples.

3.2.1 Mean Field Approximation and Self-Consistent Field

The essence of the mean field approximation is the decoupling and pre-averaging of some of the variables in the many-body correlations.

In a concentrated polymer system such as a semi-dilute polymer solution or a polymer melt, chains overlap each other. In such a concentrated system, the excluded volume interaction is canceled through screening effects (see the box entitled *Screening Effects in Concentrated Polymer Systems* on p. 55). Due to this screening effect, the correlations between different chains are suppressed and the chain conformations become almost independent of each other. Let us focus on a single chain (hereafter referred to as the tagged chain) in the system, and calculate its conformational probability distribution. A reasonable approximation in this case is the mean field approximation, where the correlations between the tagged chain and the other chains are neglected. Then the conformation of the tagged chain can be approximated as an ideal chain in an average external potential field $V(\boldsymbol{r})$ generated by the surrounding chains (see Fig. 3.5). Such an external potential field is called the mean field and will later be called the self-consistent field.

This mean field accounts for the interactions between a segment of the tagged chain (hereafter called the tagged segment) and the surrounding segments, including the other segments of the tagged chain. As such an interaction depends on the species of the tagged segment, so does the mean field $V(\boldsymbol{r})$. If we consider a system composed of several segment species $A, B, \ldots$ (an example is a blend of several types of homopolymer or a melt of a block copolymer), the mean field acting on the tagged segment of K-type at position $\boldsymbol{r}$ is given by

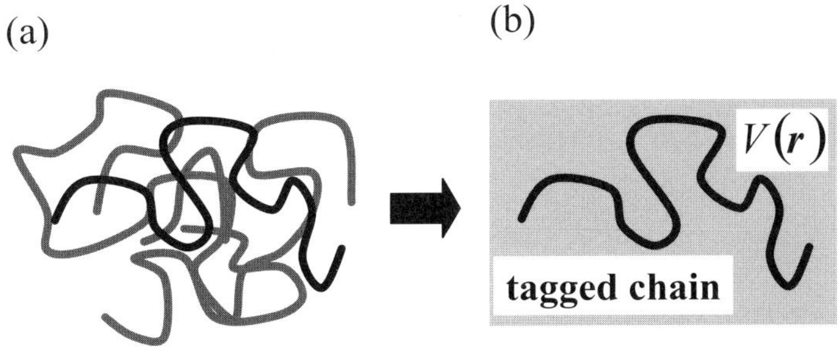

Fig. 3.5. An illustrative explanation of the mean field approximation. (**a**) An interacting many-chain system is approximated by (**b**) a single ideal chain in an averaged potential field $V(\boldsymbol{r})$

$$V_K(\boldsymbol{r}) = k_\mathrm{B}T \sum_{K'} \chi_{KK'}\phi_{K'}(\boldsymbol{r}) + \gamma(\boldsymbol{r}) \,, \tag{3.1}$$

where $\chi_{KK'}$ is the interaction parameter between a pair from a K-type segment and a K'-type segment defined by (2.68), $\phi_K(\boldsymbol{r})$ is the number density of the K-type segment at position $\boldsymbol{r}$ (number of segments per unit volume),[1] and $\gamma(\boldsymbol{r})$ is the constraint force due to external conditions such as the incompressibility condition. [We have assumed here that the constraining condition, like the incompressibility condition, does not depend on the segment species. In such a case, the constraint force $\gamma(\boldsymbol{r})$ does not depend on the segment species K.]

Our final goal is to calculate the equilibrium segment distribution function $\{\phi_K(\boldsymbol{r})\}$ taking the chain conformations into account. Such a segment density can be calculated from the equilibrium probability distribution of the tagged chain, which is assumed to be an ideal chain, in the mean field $V_K(\boldsymbol{r})$. Due to the consistency of the whole theoretical framework, the calculated segment density should be equal to the segment density appearing on the right-hand side of (3.1). After such a self-consistent condition, the mean field $\{V_K(\boldsymbol{r})\}$ is called the self-consistent field (SCF), and this theoretical framework is called the self-consistent field theory. Within this SCF theory, the equilibrium segment densities, the probability distributions of the chain conformation, and the self-consistent fields are determined simultaneously.

3.2.2 Path Integral Formalism for Polymers

In the self-consistent field theory, it is essential to calculate the equilibrium conformation of the tagged chain in the self-consistent field, which is given formally by (2.32). However, a distribution function of the chain conformation such as (2.32) is a function in $3(N+1)$-dimensional configuration space. For practical reasons, it is essential to rewrite the probability distribution in a more tractable form. Such an expression can be obtained using the path integral representation.

As shown in Fig. 3.6, let us consider part of a chain composed of K-type segments with indices $(k, k+1, \ldots, l-1, l)$. Hereafter we call it a subchain. The total number of segments participating in this subchain is $|l-k+1|$.

This subchain is assumed to be in equilibrium in the external self-consistent field $V_K(\boldsymbol{r})$, which depends on the position $\boldsymbol{r}$ and the segment species K. Then the statistical weight of this subchain having a certain conformation is given by the Boltzmann factor of the Hamiltonian (2.30) by

$$\exp\Big(-\beta\mathcal{H}\Big) = \exp\left[-\frac{3}{2b^2}\sum_{i=k}^{l-1} |\boldsymbol{r}_{i+1} - \boldsymbol{r}_i|^2 - \beta\sum_{i=k}^{l} V_K(\boldsymbol{r}_i)\right] \,, \tag{3.2}$$

[1] In Chap. 2, we used the volume fraction to describe the segment distribution using a lattice picture. In this chapter, we use a continuum description, where the volume fraction per lattice site is replaced by the number density.

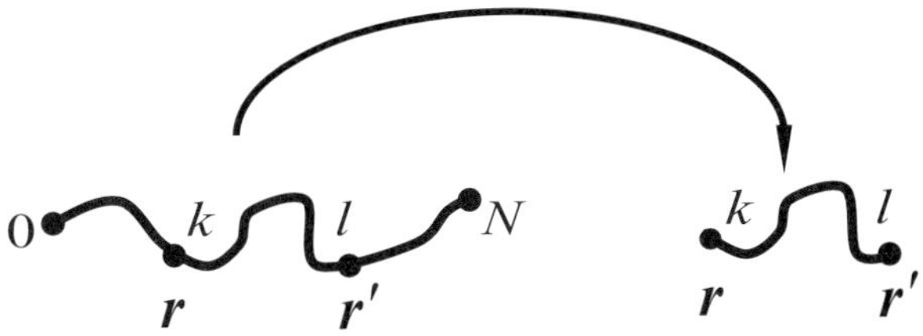

Fig. 3.6. A subchain $i \in \{k, k+1, \ldots, l-1, l\}$ and the whole chain

where we have substituted the interaction between the tagged segment and the self-consistent field into the potential $\mathcal{W}$ in (2.30). (Note that this Boltzmann factor is not the occurrence probability of the conformation of the subchain. The actual occurrence probability is proportional to a product of this quantity with the statistical weights of the other subchains connected to this subchain.) If the positions of both ends of the subchain, $i = k$ and $i = l$, are fixed at $\boldsymbol{r}$ and $\boldsymbol{r}'$, the statistical weight of such a subchain is obtained by summing the Boltzmann factor (3.2) for all possible conformations under the condition that both ends are fixed. Therefore, the total statistical weight of a subchain whose ends are fixed at $\boldsymbol{r}$ and $\boldsymbol{r}'$ is expressed in path integral form by

$$Q_K(k, \boldsymbol{r}; l, \boldsymbol{r}') \equiv \frac{\mathcal{V}}{Z_0(|l-k|)} \times \sum_{\substack{\text{all possible conformations} \\ \text{satisfying } \boldsymbol{r}_k = \boldsymbol{r} \text{ and } \boldsymbol{r}_l = \boldsymbol{r}'}} \exp\left[-\frac{3}{2b^2} \sum_{i=k}^{l-1} |\boldsymbol{r}_{i+1} - \boldsymbol{r}_i|^2 - \beta \sum_{i=k}^{l} V_K(\boldsymbol{r}_i) \right] , \tag{3.3}$$

where $\mathcal{V}$ is the system volume which accounts for the translational degree of freedom of the center of mass of the chain and $Z_0(|l-k|)$ is the partition function of an ideal subchain composed of $|l-k|$ segments defined by (2.35). [While the numerator of (3.3) contains the mean field $V_K(\boldsymbol{r})$, the denominator does not. Thus the denominator Z_0 is not the normalization constant of the probability distribution.]

From the definition (3.3) of the path integral, we can easily check the relation

$$Q_K(s, \boldsymbol{r}; s', \boldsymbol{r}') = \int \mathrm{d}\boldsymbol{r}''\, Q_K(s, \boldsymbol{r}; s'', \boldsymbol{r}'') Q_K(s'', \boldsymbol{r}''; s', \boldsymbol{r}') \,. \tag{3.4}$$

This relation expresses the statistical independence of the conformations of subchains that have no common segments (except for the end segment). It is known as the Chapman–Kolmogorov relation.

Recurrence Formula for the Path Integral

A more tractable form of the path integral can be obtained by rewriting (3.3) as a recurrence formula. Let us split the subchain $i = k, \ldots, l$ into two parts, i.e., $i = k, \ldots, l-1$ and $i = l-1, l$, as shown in Fig. 3.7. Using the Chapman–Kolmogorov relation, (3.3) is rewritten in the form

$$Q_K(k,\boldsymbol{r};l,\boldsymbol{r}') = \frac{\mathcal{V}}{Z_0(1)} \int \mathrm{d}\boldsymbol{r}'' Q_K(k,\boldsymbol{r};l-1,\boldsymbol{r}'') \exp\left[-\frac{3}{2b^2}|\boldsymbol{r}''-\boldsymbol{r}'|^2 - \beta V_K(\boldsymbol{r}')\right]. \tag{3.5}$$

Expanding the quantity $Q(k,\boldsymbol{r};l-1,\boldsymbol{r}'')$ on the right-hand side of (3.5) with respect to $\boldsymbol{r}''$ around $\boldsymbol{r}'' = \boldsymbol{r}'$, and assuming that $V_K(\boldsymbol{r})$ and its derivative are slowly varying functions with small amplitudes, we obtain

$$\begin{aligned} Q_K(k,\boldsymbol{r};l,\boldsymbol{r}') = & \left[1-(\Delta i)\beta V_K(\boldsymbol{r}')\right] \int \mathrm{d}\boldsymbol{r}'' \left(\frac{3}{2\pi(\Delta i)b^2}\right)^{3/2} \exp\left(-\frac{3}{2(\Delta i)b^2}|\boldsymbol{r}''-\boldsymbol{r}'|^2\right) \\ & \times \Big[Q_K(k,\boldsymbol{r};l-1,\boldsymbol{r}') + \nabla' Q_K(k,\boldsymbol{r};l-1,\boldsymbol{r}')\cdot(\boldsymbol{r}''-\boldsymbol{r}') \\ & + \frac{1}{2}\nabla'\nabla' Q_K(k,\boldsymbol{r};l-1,\boldsymbol{r}') : (\boldsymbol{r}''-\boldsymbol{r}')(\boldsymbol{r}''-\boldsymbol{r}') \Big], \end{aligned} \tag{3.6}$$

where we have introduced the quantity $\Delta i (\equiv 1)$, which is the increment in the index i per segment. Here, $\boldsymbol{ab}$ denotes the dyadic product of two vectors $\boldsymbol{a} = (a_i)$ and $\boldsymbol{b} = (b_i)$ defined as $\boldsymbol{ab} = (a_i b_j)$, and the notation $\boldsymbol{A} : \boldsymbol{B}$ is a scalar product of two second rank tensors $\boldsymbol{A}$ and $\boldsymbol{B}$ defined by $\boldsymbol{A} : \boldsymbol{B} \equiv \mathrm{trace}(\boldsymbol{A} \cdot \boldsymbol{B}) = \sum_{ij} A_{ij} B_{ji}$.

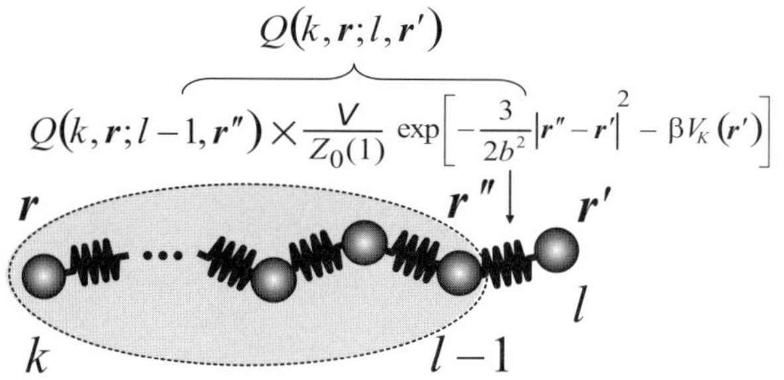

Fig. 3.7. In order to derive a recurrence relation for the path integral, the last segment is treated separately from the other segments of the subchain. Due to the statistical independence between different subchains, the orientation of the bond that connects the separated segment and the others is statistically independent of the other bonds

Quantum Mechanics and Polymer Physics

In quantum mechanics, the probability of finding a particle at time t_b at position x_b provided that the particle was found at time t_a ($t_a < t_b$) at position x_a is given by the square of the probability amplitude (wave function) $K(b,a)$ defined by [22, 23]

$$K(b,a) = \sum_{\text{all possible paths } x(t)} \exp\left\{\frac{i}{\hbar}S[x(t)]\right\} , \tag{3.7}$$

where $\hbar$ is the Planck constant divided by 2π and $S[x(t)]$ is the action of the particle. [In the same manner as the derivation of (3.10), it is easy to show that this probability amplitude $K(b,a)$ obeys the Schrödinger equation.] Here, the action $S[x(t)]$ for a one-dimensional particle with mass m moving in a potential field $V(x)$ is given by

$$S[x(t)] = \int_{t_a}^{t_b} \left[\frac{m}{2}\dot{x}^2 - V(x,t)\right] \mathrm{d}t , \tag{3.8}$$

where $\dot{x} = \mathrm{d}x/\mathrm{d}t$ is the particle velocity. Replacing $\dot{x}$ with its finite difference expression $\dot{x} \sim \left[x(t+\Delta) - x(t)\right]/\Delta$, the first and the second terms on the right-hand side of (3.8) can be identified with the spring potential and the self-consistent field of the bead–spring model given by (3.2), where a negative sign should be attached to the potential. The path integral Q_K defined in (3.3) is thus similar to the wave function of a particle in a potential field $-V(x)$. In fact, this type of correspondence is used in simulations of quantum systems, such as a solvated electron in a classical molecular fluid [20]. In such a simulation, a single quantum degree of freedom is replaced by a closed chain of classical particles, using the so-called Trotter formula [20].

We now do the integration with respect to $\boldsymbol{r}''$ on the right-hand side of (3.6). This integration can be performed using the formulas [see (2.23)]

$$\int \mathrm{d}\boldsymbol{r}'' \left(\frac{3}{2\pi(\Delta i)b^2}\right)^{3/2} \exp\left(-\frac{3}{2(\Delta i)b^2}|\boldsymbol{r}''-\boldsymbol{r}'|^2\right)(\boldsymbol{r}''-\boldsymbol{r}') = \boldsymbol{0} ,$$
$$\int \mathrm{d}\boldsymbol{r}'' \left(\frac{3}{2\pi(\Delta i)b^2}\right)^{3/2} \exp\left(-\frac{3}{2(\Delta i)b^2}|\boldsymbol{r}''-\boldsymbol{r}'|^2\right)(\boldsymbol{r}''-\boldsymbol{r}')(\boldsymbol{r}''-\boldsymbol{r}') = \frac{b^2}{3}(\Delta i)\boldsymbol{1} , \tag{3.9}$$

where $\boldsymbol{1}$ is the unit tensor. The factor Δi on the right-hand side of the second equation comes from the scaling nature of the Gaussian chain described in (2.9). After this integration, we take a continuum limit where the discrete index i is replaced by a continuous variable s (i.e., the limit $i \to s$ and

$\Delta i \to \Delta s$). Then, taking the limit $\Delta s \to 0$, we finally obtain the following diffusion-type partial differential equation:

$$\frac{\partial}{\partial s} Q_K(s', \boldsymbol{r}'; s, \boldsymbol{r}) = \left[\frac{b^2}{6} \nabla^2 - \beta V_K(\boldsymbol{r}) \right] Q_K(s', \boldsymbol{r}'; s, \boldsymbol{r}) \;, \tag{3.10}$$

where the term containing $V_K \nabla^2 Q_K$ was dropped because it is of higher order in Δs, which vanishes in the limit $\Delta s \to 0$.

Equation (3.10) is the basic equation for calculating the path integral. It is worth noting the similarity between the present formulation and quantum mechanics. If we identify the path integral Q_K with the wave function, (3.10) can be regarded as the Schrödinger equation of quantum mechanics. There is therefore a formal correspondence between polymer physics and quantum mechanics. We can import various techniques from quantum physics to polymer physics (see the box entitled *Quantum Mechanics and Polymer Physics*). A particular example using such a correspondence will be presented in Sect. 3.2.3.

Self-Consistency Condition

For a given external field $V_K(\boldsymbol{r})$, one can solve the partial differential equation (3.10) and obtain the path integral $Q_K(0, \boldsymbol{r}_0; s, \boldsymbol{r})$. Using this path integral, the density distribution of the K-type segment is calculated in the following way. As shown in Fig. 3.8, the statistical weight of a chain whose s th segment is located at the position $\boldsymbol{r}$ is calculated by integrating the product of the statistical weights of the two subchains meeting at the s th segment over all possible positions of the two end segments $\boldsymbol{r}_0$ and $\boldsymbol{r}_{N_K}$. Then the density distribution of the s th segment is given by properly normalizing this statistical weight. Integrating this distribution of the s th segment over the segment index s, we obtain the following formula for the total segment density of the K-type chain:

$$\phi_K(\boldsymbol{r}) = M_K \frac{\displaystyle\int_0^{N_K} \mathrm{d}s \int \mathrm{d}\boldsymbol{r}_0 \int \mathrm{d}\boldsymbol{r}_{N_K} Q_K(0, \boldsymbol{r}_0; s, \boldsymbol{r}) Q_K(s, \boldsymbol{r}; N_K, \boldsymbol{r}_{N_K})}{\displaystyle\int \mathrm{d}\boldsymbol{r}_0 \int \mathrm{d}\boldsymbol{r}_{N_K} Q_K(0, \boldsymbol{r}_0; N_K, \boldsymbol{r}_{N_K})} \;, \tag{3.11}$$

where M_K is the total number of K-type chains in the system and a single chain is composed of $N_K + 1$ segments. The denominator is the normalizing constant, where the normalization condition for $\phi_K(\boldsymbol{r})$ is given by

$$\int \phi_K(\boldsymbol{r}) \mathrm{d}\boldsymbol{r} = (N_K + 1) M_K \;. \tag{3.12}$$

In the denominator of (3.11), we used the Chapman–Kolmogorov relation (3.4).

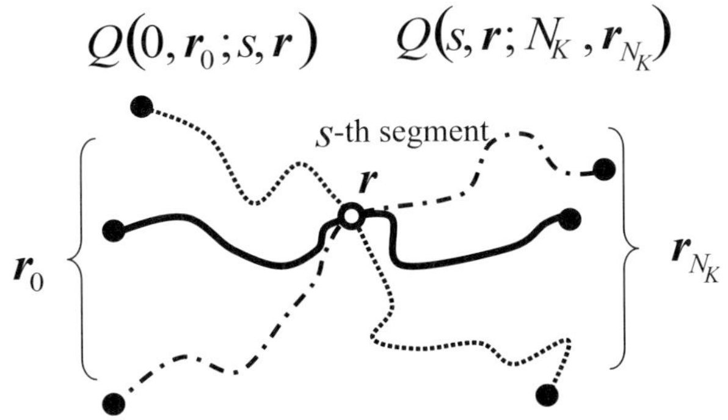

Fig. 3.8. Calculating the segment density distribution using path integrals. The density distribution of the s th segment at $\boldsymbol{r}$ is obtained by integrating the product of the path integrals for the two subchains over all possible positions of the two end segments $\boldsymbol{r}_0$ and $\boldsymbol{r}_{N_K}$ under the condition that the position of the s th segment is fixed at $\boldsymbol{r}$

The self-consistent field $\{V_K(\boldsymbol{r})\}$ should be determined so that the density distribution $\{\phi_K(\boldsymbol{r})\}$ obtained by (3.11) coincides with that assumed in (3.1). This self-consistency condition is the origin of the name of the self-consistent field. In general, the self-consistency condition is realized by a numerical iteration procedure, whose details will be described in Sect. 3.3.3.

The self-consistent field theory is useful for calculating mesoscopic structures in inhomogeneous systems, including phase separation structures in polymer blends or block copolymer melts. On the other hand, the theory gives a trivial result for homogeneous systems (see the box entitled *Path Integrals for Homogeneous Systems*).

3.2.3 Classical Approximation for Self-Consistent Field Theory

As it is generally difficult to obtain an analytical solution for the set of self-consistent equations (3.1), (3.10) and (3.11), we have to rely on numerical calculations. However, there are a few exceptions where the set of self-consistent equations can be solved analytically. One such example is found in the so-called classical limit. Here, we briefly describe the analytical treatment of the self-consistent field equations in the classical limit because such an analytic solution is important in obtaining an intuitive understanding of the self-consistent field calculation.

Path Integrals for Homogeneous Systems

In a homogeneous system, such as a single-component polymer melt in its equilibrium state, the translational invariance of the system requires the self-consistent field $\{V_K(\boldsymbol{r})\}$ to be everywhere constant. In this case, the evolution equation (3.10) for the path integral reduces to the diffusion equation of a particle in a constant external potential. As such a constant potential does not affect the motion of the diffusing particle at all, the solution $Q_K(s', \boldsymbol{r}'; s, \boldsymbol{r})$ of (3.10) is a Gaussian distribution, which means that the chain behaves as an ideal chain, irrespective of its surrounding environment. This situation is the same as is assumed in the Flory–Huggins theory described in Sect. 2.3.4. Therefore, the self-consistent field theory for a homogeneous system reduces to the Flory–Huggins mean field theory.

This characteristic can be proved mathematically in the following way. Let us add a constant term $V_K^{(0)}$ to the self-consistent field $V_K(\boldsymbol{r})$ to give $V_K(\boldsymbol{r}) + V_K^{(0)}$. Applying the transformation

$$Q_K(s', \boldsymbol{r}'; s, \boldsymbol{r}) = \tilde{Q}_K(s', \boldsymbol{r}'; s, \boldsymbol{r}) \exp\left[-\beta(s - s')V_K^{(0)}\right]$$

to the evolution equation (3.10) and the expression for the segment density (3.11), one can easily confirm that these equations are invariant. This means that a constant term in the self-consistent field can be eliminated without loss of generality. Therefore, the self-consistent field for a uniform system $V_K(\boldsymbol{r}) =$ constant is equivalent to $V_K(\boldsymbol{r}) \equiv 0$, which proves the above claims.

Classical Approximation of Path Integral

In Sect. 3.2.2, we pointed out that there is a similarity between a polymer system and a quantum system. Indeed, the path integral $Q(0, \boldsymbol{r}_0; s, \boldsymbol{r})$ for a polymer chain can be identified with the wave function of a quantum particle when we identify the segment index s with the time variable t. The distribution of each segment is then identified with the probability distribution of a quantum particle at each instant of time.

In the classical limit of such a quantum system [$h \to 0$ in (3.7)], the solution of the path integral equation (3.10) is well approximated by the classical path of Newton's equation of motion, where the fluctuations around the classical path of the particle are negligible [22]. This corresponds to a strongly stretched chain whose conformation fluctuates only weakly around the straight conformation. For a schematic illustration of the behavior of the chain in the classical limit, see Fig. 3.9.

One of the appropriate targets of this classical approximation is the polymer brush, i.e., polymer chains grafted onto a solid surface (see Fig. 3.10). For simplicity, we only consider polymer chains which are fixed on the surface

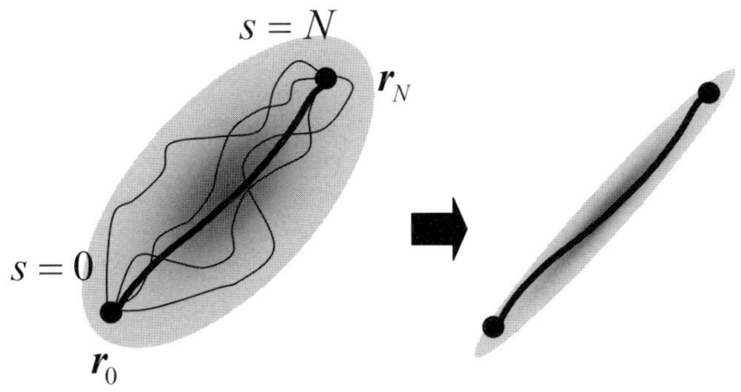

Fig. 3.9. Schematic illustration of the path integral of a chain starting from $\boldsymbol{r}_0$ at $s = 0$ and ending at $\boldsymbol{r}_N$ at $s = N$ (*left*), and its classical limit (*right*). The *bold curve* is the classical path. The actual path (*thin curves*) fluctuates around this classical path

by chemical bonds that can neither be detached nor migrate on the surface (chemically-adsorbed polymers). (Polymers adsorbed onto a solid surface by a weak molecular interaction, such as the Lennard–Jones interaction, can escape from the surface or migrate on the surface. Such polymers are called physically-adsorbed polymers.)

An important parameter that determines the physical properties of a polymer brush is the grafting density σ_a, which is defined by $\sigma_a = M/S$, where M is the number of chains grafted to the surface and S is the total area of the surface. If the average distance between neighboring grafting points is much smaller than the gyration radius R_{G} of the chain defined by (2.10), the chains overlap one another. This condition is expressed in terms of σ_a by

$$\sigma_a^{-1/2} \ll R_{\mathrm{G}} \,. \tag{3.13}$$

In this case, the chains are strongly stretched in the direction perpendicular to the solid surface due to the repulsive excluded volume interaction between the segments (see Fig. 3.10).

Let us consider the behavior of the path integral given by (3.3) in the high grafting density limit, (3.13). To understand the behavior of the path integral, it is useful to notice the correspondence between conformations of polymer chains and paths of 'quantum particles', where the segment index s is identified with the time (see the box entitled *Quantum Mechanics and Polymer Physics* on p. 109). Due to this correspondence, the first and second terms in the argument of the exponential function in (3.3) are identified with the kinetic energy and the potential energy of the 'quantum particle', respectively. The quantum uncertainty effect causes the path of the 'quantum particle' to fluctuate around its classical trajectory. This fluctuation corresponds to the thermal fluctuation of the polymer chain conformation.

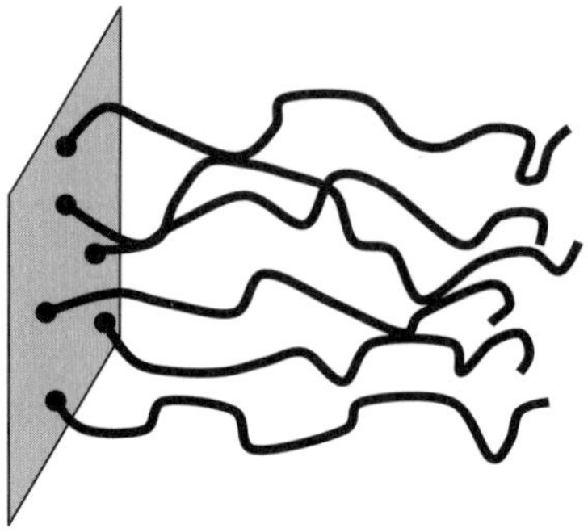

Fig. 3.10. A polymer brush grafted onto a solid surface. When the density of the grafted chains is large, the chains are stretched in the direction perpendicular to the solid surface due to the excluded volume interaction

In the limit of (3.13), the chain conformation becomes almost straight due to the strong excluded volume interaction. As a result, the fluctuation (uncertainty) in the trajectory of the 'quantum particle' becomes less important. Then the trajectory of the 'quantum particle' converges to a certain unique path, which corresponds to the classical path. Such a classical path $\{\boldsymbol{r}_i\}$ is defined as the one that maximizes the argument of the exponential function on the right-hand side of (3.3). This extremal condition leads to the following equation for arbitrary segment index i:

$$\frac{\partial}{\partial z_i}\left[-\frac{1}{2b^2}\sum_j |z_j - z_{j+1}|^2 - \beta \sum_j V(z_j)\right] = 0\,, \tag{3.14}$$

where the z-axis is taken in the direction perpendicular to the solid surface, z_i is the z-component of $\boldsymbol{r}_i$, and we assume that the system is uniform in the direction parallel to the solid surface. Because of the one-dimensional nature of the problem, the coefficient of the spring force is no longer $3/(2b^2)$ but $1/(2b^2)$. Performing the partial differentiation with respect to z_i and taking the continuum limit with respect to i, we obtain

$$\frac{k_{\mathrm{B}}T}{b^2}\frac{\mathrm{d}^2}{\mathrm{d}s^2}z(s) = V'\big(z(s)\big)\,. \tag{3.15}$$

If we identify the segment index s as time, (3.15) corresponds to Newton's equation of motion for a classical particle with mass $k_{\mathrm{B}}T/b^2$ moving in a potential $-V(z)$.

Applications of Polymer Brushes

The polymer brushes studied in this section can be applied to various industrial problems. Grafting polymer chains onto a solid surface or onto surfaces of colloidal particles alters the nature of the interaction between these surfaces.

When colloidal particles are dissolved in a dilute polymer solution, these particles easily form aggregates because an effective attractive interaction is induced between the particles by the polymers dissolved in the solvent. This phenomenon is called the depletion effect. The polymers are excluded from the neighborhood of the surfaces of the colloidal particles in order to avoid loss of conformational entropy of the chains associated with the deformation due to the surface of the particle. As a result, the polymer density is reduced within a layered region with width R_{G} around the surface of the colloidal particle (where R_{G} is the gyration radius of the chain). This layer is called the depletion layer.

The existence of the depletion layer reduces the volume available for occupation by the centers of mass of the polymer chains by an amount $4\pi a^2 R_{\mathrm{G}}$ per colloidal particle, where a is the radius of the colloidal particle. This reduction in the free volume corresponds to a loss of the translational entropy of the polymer chains. When two colloidal particles approach and their depletion layers overlap, the free volume for the polymers is increased, and this leads to a decrease in the total free energy of the system. This is equivalent to an effective attractive interaction between the two colloidal particles. Due to this attractive interaction, the colloidal particles tend to aggregate in dilute polymer solutions. However, if the colloidal particles are covered by polymer brushes, such an attractive interaction does not take place because the depletion regions are already occupied by the grafted polymers. Instead, an overlapping of the grafted polymer brushes generates a repulsive interaction when two colloidal particles approach. Therefore, the polymer-grafted colloidal particles are stable against aggregation even in dilute polymer solutions.

Classical Approximation for Polymer Brushes

There are two typical limiting cases for a polymer brush in a solvent. One is a brush subjected to a good solvent that has affinity with the polymer segments. In this case, the solvent can permeate into the brush and swell it. Such a brush is called a swollen brush. On the other hand, if the solvent is a poor solvent that dislikes the polymer segments, the solvent molecules are excluded from the brush. In this case, the brush is in a melt state of the polymer, and is called a melt brush (see Fig. 3.11).

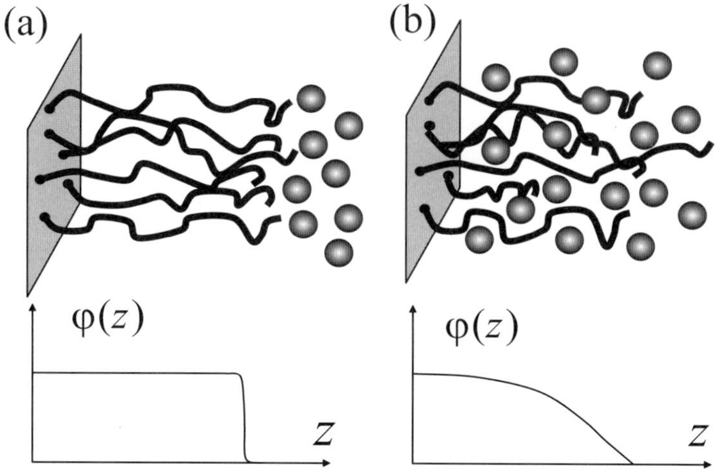

Fig. 3.11. Two typical states of polymer brushes, i.e., (**a**) the melt brush where the solvent (*open circles*) is excluded from the brush, and (**b**) the swollen brush where the solvent can permeate into the brush. *Lower pictures* show the density profiles of the segments in the direction perpendicular to the grafting surface

Let us solve (3.15) for the swollen brush and the melt brush, respectively. For simplicity, we assume that the grafting surface is flat and that all chains are the same type of homopolymer composed of N segments.

The most important characteristic of the brush is that one end of the chain is grafted onto the solid surface at $z = 0$. Let us assume that this end segment is specified by $s = N$. We then obtain the constraint condition

$$z(N) = 0 \; . \tag{3.16}$$

This is a boundary condition for (3.15) at $s = N$. On the other hand, the other end segment $s = 0$ is free, and is called the free end. In this case, the boundary condition at $s = 0$ is derived from the vanishing tension at the free end. As we assume a bead–spring model, the tension in the chain is proportional to the stretching of the chain $\mathrm{d}z/\mathrm{d}s$. Then the boundary condition at $s = 0$ is given by

$$z'(0) = 0 \; , \tag{3.17}$$

where the prime denotes differentiation with respect to the argument.

The classical trajectory corresponds to the solution $z(s)$ of (3.15) that starts from a rest state $[z'(0) = 0]$ at position $z(0)$ at time $s = 0$ and arrives at $z = 0$ at time $s = N$. The characteristic property of this problem is that the particle has to arrive at $z = 0$ at the time $s = N$ irrespective of its initial position $z(0)$. Such a feature is realized by adopting the isochronisms of a pendulum, which means that the period of a pendulum is always the same irrespective of its amplitude as long as the amplitude is small. Using this feature, we understand that the appropriate potential $V(z)$ for our brush is given by the harmonic potential

$$V(z) = \alpha\left(h^2 - z^2\right) , \tag{3.18}$$

where α is a constant parameter and h is the height of the upper bound of the brush measured from the grafting surface. As the self-consistent field has an undetermined constant contribution (see the box entitled *Path Integrals for Homogeneous Systems* on p. 112), we can assume without loss of generality that the self-consistent field $V(z)$ vanishes at $z = h$. Using this potential, the solution of (3.15) is given by

$$z(s) = z_0 \cos\frac{\pi s}{2N} , \tag{3.19}$$

where $z_0 = z(0)$ is the position of the free end. The consistency condition between the potential (3.18) and the solution given by (3.19) requires that the constant α in (3.18) be given by

$$\alpha = \frac{\pi^2 k_{\mathrm{B}} T}{8N^2 b^2} . \tag{3.20}$$

Next we calculate the probability distribution of the sth segment of a chain whose free end is located at position z_0. Let us denote this distribution by $\phi(z|z_0)$. This probability is inversely proportional to the speed of the classical particle whose path is given by (3.19), and is given by

$$\phi(z|z_0) = \sigma_{\mathrm{a}}\left|\frac{\mathrm{d}}{\mathrm{d}s}z(s)\right|^{-1} = \frac{2N\sigma_{\mathrm{a}}}{\pi}\frac{1}{\sqrt{z_0^2 - z^2}} , \tag{3.21}$$

where σ_{a} is the grafting density, defined as the number of grafted chains per unit area of the surface. If the probability distribution of the free end $g(z_0)$ is given, the total segment density is obtained by

$$\phi(z) = \int_z^h \mathrm{d}z_0\, g(z_0)\phi(z|z_0) = \frac{2N\sigma_{\mathrm{a}}}{\pi}\int_z^h \frac{g(z_0)}{\sqrt{z_0^2 - z^2}}\mathrm{d}z_0 , \tag{3.22}$$

which means that $\phi(z)$ and $g(z_0)$ are related to each other through this integral transformation.

Structure of Polymer Brushes in the Classical Approximation

Equations (3.19) and (3.22) can be applied to both swollen brushes and melt brushes. The difference between these two types of brush arises from the difference in the distribution function of the free ends $g(z_0)$ [24]:

- **Swollen Brush Case.** As there is no constraint on the segment density, the self-consistent field $V(z)$ is determined only by the contribution from the excluded volume interaction. Using the excluded volume parameter v and the segment density $\phi(z)$, it is described by

$$V(z) = k_{\mathrm{B}}Tv\phi(z) \qquad \text{(swollen brush)}\ , \tag{3.23}$$

which means that the profile of the segment density $\phi(z)$ is a parabolic function of z [see (3.18)]. In this case, the free end density $g(z_0)$ in (3.22) are given by

$$g(z_0) = \frac{3}{h^3} z_0 \left(h^2 - z_0^2\right)^{1/2} . \tag{3.24}$$

The brush height h is determined by the relation

$$h = \left(\frac{12b^2\sigma_{\mathrm{a}}v}{\pi^2}\right)^{1/3} N . \tag{3.25}$$

- **Melt Brush Case.** In this case, the total segment density should be constant at any point in the brush (incompressibility condition). Due to this constraint condition, the self-consistent field is given by

$$V(z) = \gamma(z) \qquad \text{(melt brush)}\ , \tag{3.26}$$

where $\gamma(z)$ is the Lagrange multiplier for the incompressibility condition, which corresponds to the pressure against compression. We then find that

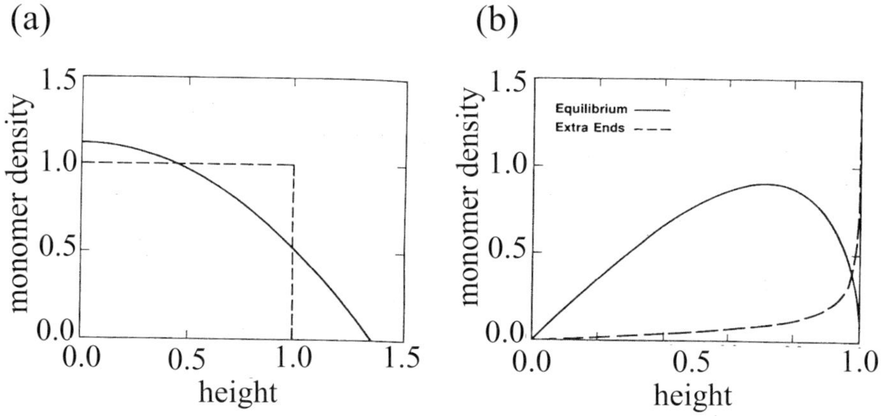

Fig. 3.12. Density profiles of a swollen brush (*continuous curve*) and a melt brush (*broken curve*) obtained using the classical approximation. (**a**) Segment density distribution $\phi(z)$ and (**b**) distribution of the free ends $g(z_0)$. The density of the free ends for the melt brush case is the sum of the *continuous curve* and the *broken curve*. Reprinted with permission from S.T. Milner et al., Macromolecules **21**, 2610 (1988). (© 1988, American Chemical Society)

$$g(z_0) = \frac{1}{h} \frac{z_0/h}{\left[1 - (z_0/h)^2\right]^{1/2}} \tag{3.27}$$

gives a self-consistent set of solutions.

The solutions for these brushes within the classical approximation are shown in Fig. 3.12 for both the swollen brush case and the melt brush case.

Integral Transformation Between Segment Density and Free End Density

The integral transformation between the segment density $\phi(z)$ and the free end density $g(z_0)$ are mutually related through an integral transformation of the following type [see (3.22)]:

$$\tilde{f}(z) = \int_z^1 \frac{f(x)}{\sqrt{x^2 - z^2}} \mathrm{d}x \, . \tag{3.28}$$

The results of this integral transformation of several typical functions are listed below:

$f(x)$	$\tilde{f}(z)$
1	$\log \left\| \frac{1 + \sqrt{1 - z^2}}{z} \right\|$
x	$\sqrt{1 - z^2}$
$x\sqrt{1 - x^2}$	$\frac{\pi}{4}(1 - z^2)$
$\frac{x}{\sqrt{1 - x^2}}$	$\frac{\pi}{2}$

In the melt brush case, the free ends of the chain tend to segregate at the free surface of the brush between the polymer and the solvent. Such an increase in the density of the free ends at the free surface of the brush is similar to the segregation of the free ends of a polymer melt at the solid surface when the polymer melt is pressed onto a solid surface. These segregation phenomena originate from the loss in translational entropy of the segments induced by the spatial constraint. The solid surface or the interface between the brush and the solvent limits the number of possible spatial arrangements of the segments. As the end segments possess more freedom than those at the middle of the chains, the end segments tend to migrate to the interfacial region.

Comparing the Classical Approximation with Other Theories

It is known that the classical approximation can give a reasonable prediction of the structure of the polymer brush. Indeed, the results of the classical approximation agree well with the experimental data obtained by neutron scattering experiments and certain other theoretical calculations, such as a direct numerical evaluation of the self-consistent field theory or molecular simulations.

Figures 3.13 and 3.14 compare the data obtained with molecular simulations (MD and MC described in Sect. 2.7) and the predictions by the classical approximation of the structure of a swollen brush grafted onto a solid surface. Figure 3.13 shows the results of MD simulations on a swollen brush. We can confirm that the chains are stretched in the direction perpendicular to the solid wall due to the excluded volume interaction, which justifies our assumption in the classical approximation for polymer brushes. As the grafting density is increased from Fig. 3.13a to Fig. 3.13c, the chains are more and more stretched.

Figure 3.14 shows the result of the segment density distribution obtained by an MC simulation using the bead–spring model along with the result of the classical approximation. The simulation result for the segment density (Fig. 3.14a) is well reproduced by the parabolic curve predicted by the classical approximation. The end segment distribution (Fig. 3.14b) can also be well fitted by the theoretical result. As a result, although the classical ap-

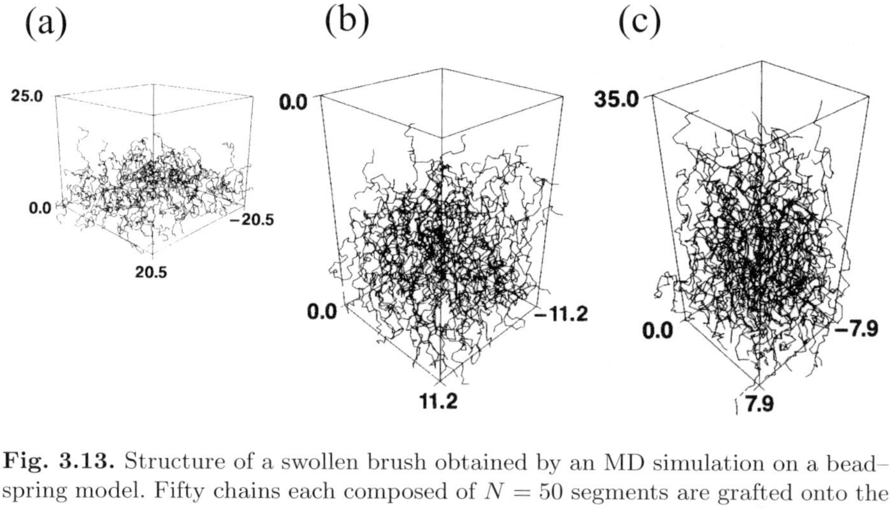

Fig. 3.13. Structure of a swollen brush obtained by an MD simulation on a bead–spring model. Fifty chains each composed of $N = 50$ segments are grafted onto the solid wall with various grafting densities σ_a. The grafting density is $\sigma_a \sigma^2 =$ (**a**) 0.03, (**b**) 0.1, and (**c**) 0.2, where σ is the diameter of the Lennard–Jones potential of the segment–segment interaction. Reprinted with permission from M. Murat and G.S. Grest, Macromolecules **22**, 4054 (1989). (© 1989, American Chemical Society)

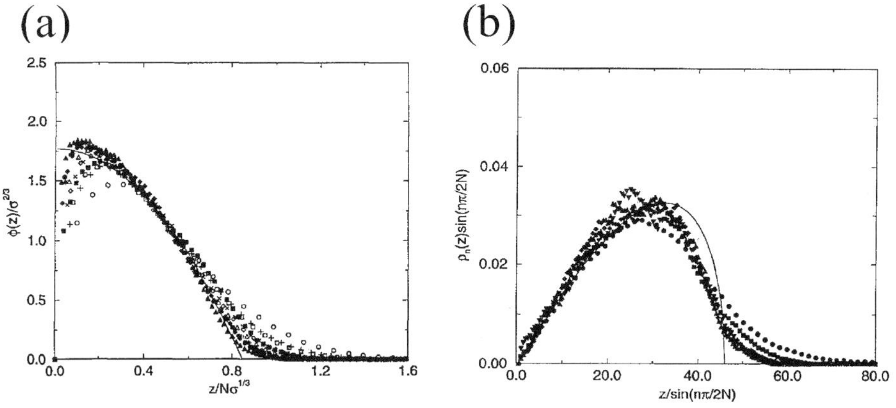

Fig. 3.14. Comparison between the results of an MC simulation and the classical approximation for the structure of a swollen brush. (**a**) Segment distribution in the direction perpendicular to the wall and (**b**) distribution of the free ends. The chain length is (**a**) $N = 40$–100, and (**b**) $N = 100$. *Symbols* and *continuous curves* are simulation results and results of the classical approximation, respectively. Reprinted with permission from M. Laradji et al., Phys. Rev. E **49**, 3199 (1994). (© 1994, American Physical Society)

proximation is simple, it can give qualitative (and sometimes quantitative) predictions of the behavior of highly stretched polymers.

3.3 Numerical Methods for the Self-Consistent Field Theory of Polymers

When combined with a numerical technique, the self-consistent field theory and the path integral formalism introduced in the previous sections can give useful and accurate predictions regarding the complex phase behavior of polymer mixtures and complex domain structures. In this section, we describe the details of the numerical method for calculating the equilibrium structures and phase diagrams of polymer mixtures and adsorbed polymers. Some examples of the applications are also shown to demonstrate the efficiency of the method.

A construction of the phase diagram requires calculation of the spatial structures and their free energy values. Among several (sometimes many) candidates for the equilibrium structures, the true equilibrium structure is the one that minimizes the free energy. We must therefore evaluate the values of the free energy for all the candidates and compare them to find the minimum free energy. In order to perform this procedure, we need an expression for the free energy of polymer mixtures within the framework of the self-consistent field theory. Before deriving such an expression, we first introduce the idea of functionals and their derivatives and integrals in Sect. 3.3.1.

Using these concepts, the general expression for the free energy is derived in Sect. 3.3.2. The actual numerical procedures to obtain the equilibrium structures and the phase diagrams are described in Sect. 3.3.3. Several examples are also described.

3.3.1 Functionals, Functional Derivatives and Functional Integrals

Definition of Functionals

In this section, we introduce the concept of functionals and their derivatives and integrals, which are necessary tools for deriving the general expression for the free energy.

While a function is a mapping from numbers to numbers, a functional is defined as a mapping from functions to numbers (see Fig. 3.15). In this book, we denote a functional F of a function $f(\boldsymbol{r})$ by $F[f(\boldsymbol{r})]$. A simple example of a functional is

$$F[f(x)] = \int_{-\infty}^{\infty} f(x)\mathrm{d}x \; , \tag{3.29}$$

which maps any function $f(x)$ to its integrated value.

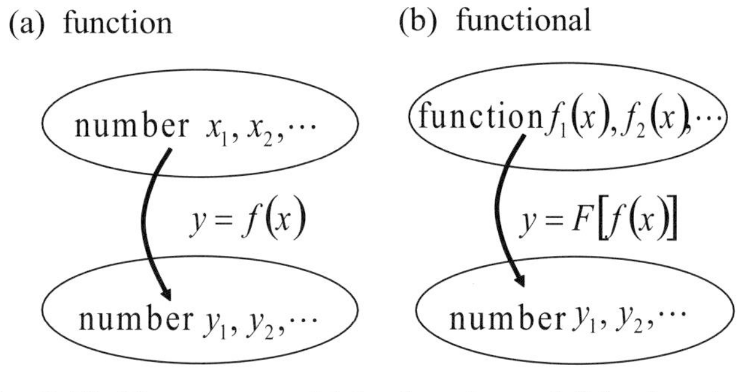

Fig. 3.15. The concepts of (**a**) a function and (**b**) a functional

Functional Derivative

Once we derive an expression for the free energy of the target system F as a functional of the segment density $\{\phi_K(\boldsymbol{r})\}$ in the form $F[\{\phi_K\}]$, the equilibrium structure is obtained as a particular segment density $\{\phi_K(\boldsymbol{r})\}$ that minimizes this free energy functional $F[\{\phi_K\}]$. The problem of finding the equilibrium state thus reduces to the problem of minimizing a functional $F[\{\phi_K\}]$ with respect to its argument $\{\phi_K(\boldsymbol{r})\}$. Such a minimization problem is formulated in terms of the derivative of $F[\{\phi_K\}]$, which is called the functional derivative.

The functional derivative is most easily defined using a discretized lattice picture of the space coordinate, as we did in Chap. 2. Let us consider a function $f(\boldsymbol{r})$ of the position $\boldsymbol{r}$ and its functional $F[f(\boldsymbol{r})]$, and define them on a spatial lattice (see Fig. 3.16). Specifying each of the lattice points by an index i as $\{\boldsymbol{r}_0, \boldsymbol{r}_1, \ldots, \boldsymbol{r}_i, \ldots, \boldsymbol{r}_L\}$, the function $f(\boldsymbol{r})$ is described by a set of values of $f(\boldsymbol{r})$ at the lattice points $\{f_0, f_1, \ldots, f_L\}$, where $f_i \equiv f(\boldsymbol{r}_i)$ is the value of $f(\boldsymbol{r})$ at the ith lattice point and L is the total number of lattice points. In this case, the functional $F[f(\boldsymbol{r})]$ is expressed as a multivariable function $F(f_0, f_1, \ldots, f_L)$.

Let us imagine changing the value f_i at the lattice point $\boldsymbol{r}_i$ by a small amount Δf_i. This will cause a change ΔF in $F(f_0, \ldots, f_L)$. In the limit $\Delta f_i \to 0$, the ratio $\Delta F/\Delta f_i$ converges to the partial derivative as

$$\frac{\Delta F}{\Delta f_i} \longrightarrow \frac{\partial F}{\partial f_i} . \tag{3.30}$$

The total variation (total difference) of F caused by the variations Δf_i for $i = 0, \ldots, L$ is given by

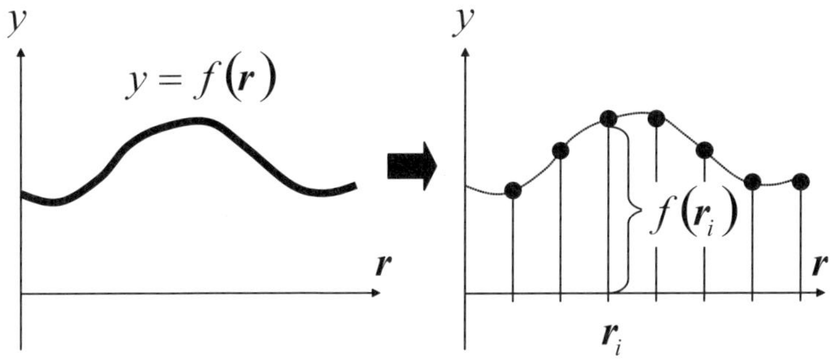

Fig. 3.16. Relation between the continuum picture (*left*) and the lattice picture (*right*)

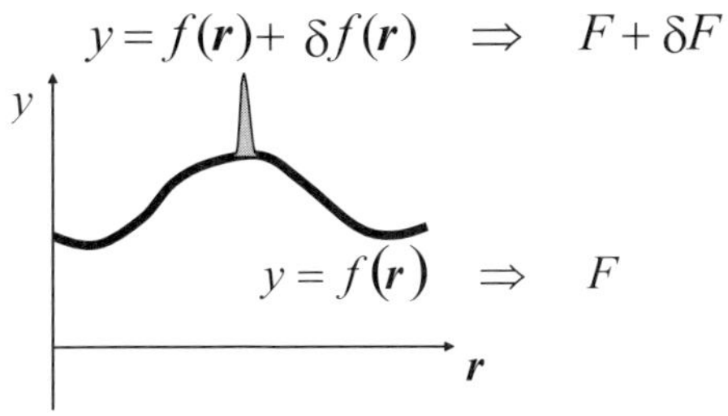

Fig. 3.17. The concept of functional derivative. When a small variation $\delta f(\boldsymbol{r})$ of the function $f(\boldsymbol{r})$ induces a variation δF of the functional $F[f(\boldsymbol{r})]$, the ratio between δF and $\delta f(\boldsymbol{r})$ defines the functional derivative $\delta F/\delta f(\boldsymbol{r})$

$$\Delta F = \sum_i \frac{\partial F}{\partial f_i} \Delta f_i \,. \tag{3.31}$$

As the lattice becomes finer and finer ($L \to \infty$), the discrete lattice turns into a continuous space and the sum $\sum_i$ over the lattice points changes to an integral over the space $\int \mathrm{d}\boldsymbol{r}$. In this limit, the discrete description (3.31) is rewritten in the continuous form

$$\Delta F = \int \mathrm{d}\boldsymbol{r} \frac{\delta F}{\delta f(\boldsymbol{r})} \Delta f(\boldsymbol{r}) \,. \tag{3.32}$$

In deriving (3.32) from (3.31), we included the volume assigned to each lattice point in the definition of $\delta F/\delta f(\boldsymbol{r})$. The coefficient $\delta F/\delta f(\boldsymbol{r})$ is defined as the functional derivative of $F[f(\boldsymbol{r})]$ with respect to $f(\boldsymbol{r})$ (see Fig. 3.17). The correspondence between (3.31) and (3.32) clearly indicates that the functional derivative is a continuum limit of partial differentiation in the discrete lattice description. For actual calculations of the functional derivatives, readers should refer to the box entitled *Useful Formulas for Functional Derivatives*.

Useful Formulas for Functional Derivatives

A useful formula for calculating the functional derivative is

$$\frac{\delta f(x)}{\delta f(y)} = \delta(x-y) \,.$$

This formula says that the variations of a function $f(x)$ at different positions x and y are mutually independent. Using this formula, we can calculate the functional derivative of an integral of an arbitrary function $g(f)$ of f as

$$\begin{aligned}
\frac{\delta}{\delta f(x)} \int_{-\infty}^{\infty} \mathrm{d}y \, g(f(y)) &= \int_{-\infty}^{\infty} \mathrm{d}y \frac{\delta}{\delta f(x)} g(f(y)) \\
&= \int_{-\infty}^{\infty} \mathrm{d}y \, g'(f(y)) \frac{\delta f(y)}{\delta f(x)} \\
&= \int_{-\infty}^{\infty} \mathrm{d}y \, g'(f(y)) \delta(x-y) = g'(f(x)) \,.
\end{aligned}$$

Similarly, we obtain

$$\begin{aligned}
\frac{\delta}{\delta f(x)} \int_{-\infty}^{\infty} \mathrm{d}y \left[\frac{\mathrm{d}f(y)}{\mathrm{d}y}\right]^2 &= \int_{-\infty}^{\infty} \mathrm{d}y \, 2 \frac{\mathrm{d}f(y)}{\mathrm{d}y} \frac{\delta}{\delta f(x)} \frac{\mathrm{d}f(y)}{\mathrm{d}y} \\
&= \int_{-\infty}^{\infty} \mathrm{d}y \, 2 \frac{\mathrm{d}f(y)}{\mathrm{d}y} \frac{\mathrm{d}}{\mathrm{d}y} \delta(x-y) \\
&= -\int_{-\infty}^{\infty} \mathrm{d}y \, 2 \frac{\mathrm{d}^2 f(y)}{\mathrm{d}y^2} \delta(x-y) = -2 \frac{\mathrm{d}^2 f(x)}{\mathrm{d}x^2} \,.
\end{aligned}$$

Functional Taylor Series Expansion

Using the concept of the functional derivative, we can also define the functional Taylor series expansion, whose definition can also be given conveniently on a lattice. As discussed previously, the discrete description of the functional $F[f(\boldsymbol{r})]$ using a lattice picture corresponds to a multivariable function $F(f_0, \ldots, f_L)$. The Taylor series expansion of this multivariable function $F(f_0, \ldots, f_L)$ around $f_i = 0$ $(i = 0, \ldots, L)$ is given by

$$F(f_0, \ldots, f_L) = F(0, \ldots, 0) + \sum_i \left. \frac{\partial F}{\partial f_i} \right|_0 f_i + \frac{1}{2} \sum_i \sum_j \left. \frac{\partial^2 F}{\partial f_i \partial f_j} \right|_0 f_i f_j + \cdots , \tag{3.33}$$

where the symbol $|_0$ means that the quantity on the left-hand side of this symbol should be evaluated for $f_i = 0$ $(i = 0, \ldots, L)$.

The continuum limit of (3.33) leads to

$$\begin{aligned} F[f(\boldsymbol{r})] = F[0] &+ \int \mathrm{d}\boldsymbol{r} \left. \frac{\delta F}{\delta f(\boldsymbol{r})} \right|_0 f(\boldsymbol{r}) \\ &+ \frac{1}{2} \int \mathrm{d}\boldsymbol{r} \int \mathrm{d}\boldsymbol{r}' \left. \frac{\delta^2 F}{\delta f(\boldsymbol{r}) \delta f(\boldsymbol{r}')} \right|_0 f(\boldsymbol{r}) f(\boldsymbol{r}') + \cdots . \end{aligned} \tag{3.34}$$

Here, the symbol $|_0$ indicates an evaluation for $f(\boldsymbol{r}) \equiv 0$. Equation (3.34) defines the functional Taylor series expansion.

Functional Integral

Finally, we define the functional integral. The functional integral of a functional $F[f(\boldsymbol{r})]$ is defined as a sum of the value of $F[f(\boldsymbol{r})]$ over all possible functions $f(\boldsymbol{r})$, i.e.,

$$\int \mathcal{D}\{f(\boldsymbol{r})\} F[f(\boldsymbol{r})] . \tag{3.35}$$

(There are several different notations for the functional integral.) Again the lattice picture can give us an intuitive understanding of this functional integral. In the lattice picture, summing the integrand over all possible functions, $f(\boldsymbol{r})$ can be replaced by a set of independent integrations with respect to the value of the function at each of the lattice points. Then the lattice description of the functional integral (3.35) is given by (see Fig. 3.18)

$$\int \mathcal{D}\{f(\boldsymbol{r})\} F[f(\boldsymbol{r})] \equiv \lim_{L \to \infty} C_{\mathrm{L}} \int_{-\infty}^{\infty} \mathrm{d}f_1 \ldots \int_{-\infty}^{\infty} \mathrm{d}f_L \, F(f_1, \ldots, f_L) , \tag{3.36}$$

where C_{L} is a normalization constant associated with the lattice structure, whose precise form is not important in the following discussion.

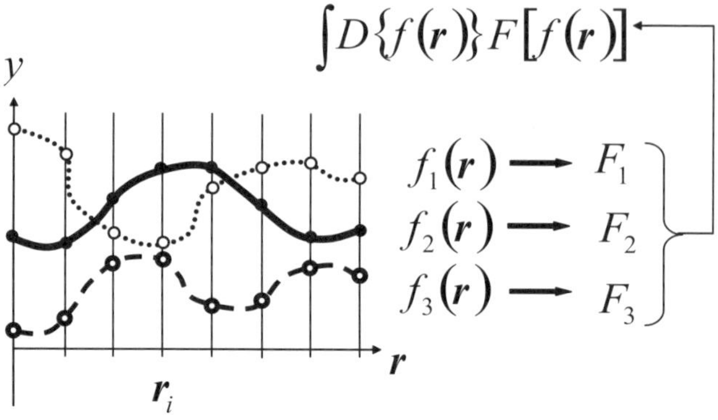

Fig. 3.18. Intuitive explanation of the functional integral

3.3.2 General Expression for the Free Energy

Free Energy Functional and Prediction of Equilibrium Structures

There are two typical methods for calculating equilibrium structures using the self-consistent field theory:

- List all the candidates for the equilibrium structure and evaluate the free energy of each. The equilibrium structure is the one that has the minimum free energy (static method).
- Simulate the time evolution of the system starting from an appropriate initial state, and observe the final structure (dynamic method).

In both methods, it is essential to obtain a general expression for the free energy of the system (see Fig. 3.19). To bring out the essence of the derivation of the free energy functional, we use here a simple one-component polymer solution, i.e., a mixture of a polymer and a solvent. Extensions to more complex situations such as multicomponent polymer mixtures will not be discussed, although the derivation is straightforward [25, 26].

General Expression for the Model Hamiltonian and Canonical Partition Function

Let us consider a polymer solution containing M linear homopolymer chains of the same type. Each of these polymer chains is composed of $N+1$ segments of the same type, numbered $i = 0, 1, \ldots, N$. As in (2.73), we assume that the degrees of freedom of the solvent are expressed in terms of the excluded volume parameter v. Under these assumptions, let us express the Helmholtz free energy F of this polymer solution as a functional of the segment density distribution $\phi(\boldsymbol{r})$.

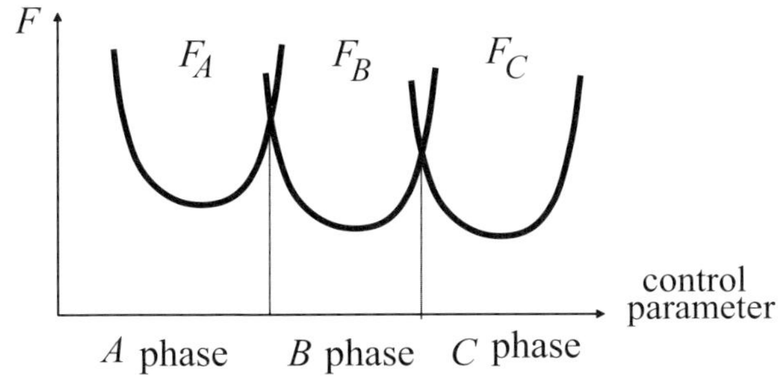

Fig. 3.19. Constructing a phase diagram by evaluating the free energy. A, B and C are three different phases, and F_A, F_B and F_C are the free energies per unit mass of these phases. (If there is a conservation law for the total number of particles, such as the phase separation of a binary alloy, we have to draw the tangent line common to the free energy curves of the two coexisting phases. Then the two contact points correspond to the coexisting states)

In the self-consistent field theory, a polymer chain interacting with the surrounding chains is modeled as an ideal chain in a mean field. (Hereafter we call this chain the tagged chain.) According to this concept, we split the total Hamiltonian $\hat{\mathcal{H}}$ of the M-chain system into two parts:

$$\hat{\mathcal{H}}(\Gamma) = \hat{\mathcal{H}}_0(\Gamma) + \hat{\mathcal{W}}(\Gamma) , \tag{3.37}$$

where $\hat{\mathcal{H}}_0$ is the Hamiltonian of the reference system composed of M independent ideal chains, and $\hat{\mathcal{W}}$ is the deviation from the ideal reference system due to the interaction. Here, Γ is the $3M(N+1)$-dimensional phase space spanned by the positions of $M \times (N+1)$ segments. (For concentrated polymer systems, the contribution from the momentum is usually negligible. In such a case, Γ should be called the configuration space.) The hat symbol indicates that the associated physical quantity is a function of the microscopic states described by the phase space Γ.

Assuming the canonical ensemble, where temperature, volume and number of particles are fixed, the partition function of this system is given by

$$Z = \frac{1}{M!} \int \mathrm{d}\Gamma \exp\Big\{ -\beta \Big[\hat{\mathcal{H}}_0(\Gamma) + \hat{\mathcal{W}}(\Gamma) \Big] \Big\} , \tag{3.38}$$

where $\int \mathrm{d}\Gamma$ is the integral over the whole phase space, expressed using the position of the ith segment of the pth chain $\boldsymbol{r}_i^{(p)}$ as

$$\mathrm{d}\Gamma \equiv \prod_{p=1}^{M} \prod_{i=0}^{N} \mathrm{d}\boldsymbol{r}_i^{(p)} . \tag{3.39}$$

The factor $1/M!$ in (3.38) is the quantum correction that originates from the fact that chains of the same type cannot be distinguished.

Coarse-Grained Segment Density Distribution

We now obtain the segment density distribution at position $\boldsymbol{r}$. From the microscopic point of view, the segment density distribution is expressed as a function of the phase point Γ by

$$\hat{\phi}(\boldsymbol{r};\Gamma) = \sum_{p=1}^{M}\sum_{i=0}^{N}\delta(\boldsymbol{r}-\boldsymbol{r}_i^{(p)}) \,. \tag{3.40}$$

This is a set of delta peaks located at the position of each of the segments. Therefore this microscopic density distribution diverges at the segment positions and vanishes elsewhere. What is necessary for us is the averaged quantity where the detailed information about the microscopic positions of the individual segments is eliminated. Such an averaged segment density distribution $\phi(\boldsymbol{r})$ is obtained by averaging the microscopic description (3.40) over the canonical ensemble generated by the partition function (3.38). This leads to

$$\begin{aligned}\phi(\boldsymbol{r}) &= \left\langle \hat{\phi}(\boldsymbol{r};\Gamma)\right\rangle \\ &= \frac{1}{M!Z}\int \mathrm{d}\Gamma\,\hat{\phi}(\boldsymbol{r};\Gamma)\exp\left\{-\beta\left[\hat{\mathcal{H}}_0(\Gamma)+\hat{\mathcal{W}}(\Gamma)\right]\right\} ,\end{aligned} \tag{3.41}$$

where $\langle\ \rangle$ is the average over the canonical ensemble derived from (3.38). Note that ϕ on the left-hand side is already averaged over the microscopic configurations of the segments and does not depend on Γ. This is why the hat symbol is not attached to ϕ on the left-hand side (see Fig. 3.20).

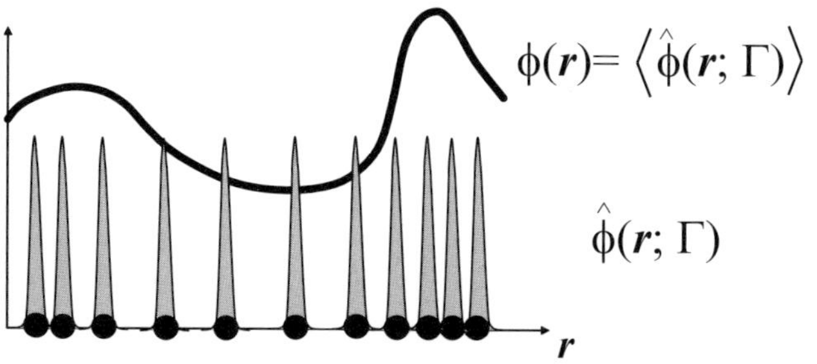

Fig. 3.20. Illustration of a microscopic description of the density distribution $\hat{\phi}(\boldsymbol{r};\Gamma)$ and its coarse-grained description $\phi(\boldsymbol{r})$

Mean Field Approximation and Steepest Descent Method

We assume that the deviation $\hat{\mathcal{W}}(\Gamma)$ from the reference state $\hat{\mathcal{H}}_0(\Gamma)$ originates from the interaction between segments, and also that $\hat{\mathcal{W}}(\Gamma)$ depends on Γ only through $\hat{\phi}(\boldsymbol{r};\Gamma)$, i.e.,

$$\hat{\mathcal{W}}(\Gamma) \equiv \hat{\mathcal{W}}[\hat{\phi}(\boldsymbol{r};\Gamma)] \,. \tag{3.42}$$

In the box entitled *Delta Functional*, we give the definition of the delta functional, which extends Dirac's delta function to functionals. Using the assumption (3.42) and the property of the delta functional (3.45), we can rewrite the expression for the partition function (3.38) in the form

$$Z = \int \mathcal{D}\{\varphi(\boldsymbol{r})\} \int \mathrm{d}\Gamma \delta[\varphi(\boldsymbol{r}) - \hat{\phi}(\boldsymbol{r};\Gamma)] \exp\Big\{-\beta\hat{\mathcal{H}}_0(\Gamma) + \mathcal{W}[\varphi(\boldsymbol{r})]\Big\} \,, \tag{3.43}$$

where $\varphi(\boldsymbol{r})$ is just a dummy function that vanishes when the functional integral $\int \mathcal{D}\{\varphi(\boldsymbol{r})\}$ is performed. Later, we shall see that this $\varphi(\boldsymbol{r})$ coincides with the average segment density $\phi(\boldsymbol{r})$ defined in (3.41), when the mean-field approximation is applied. Using the Fourier transformation (3.46) of the delta functional, we can rewrite (3.43) as

$$\delta[\phi(\boldsymbol{r})] = \int \mathcal{D}\{w(\boldsymbol{r})\} \exp\left[\mathrm{i}\int w(\boldsymbol{r})\phi(\boldsymbol{r})\mathrm{d}\boldsymbol{r}\right] \,. \tag{3.44}$$

Delta Functional

Similarly to the Dirac delta function $\delta(\boldsymbol{r})$, we can define the delta functional $\delta[f(\boldsymbol{r})]$ as

$$\int \mathcal{D}\{f(\boldsymbol{r})\}\delta[f(\boldsymbol{r}) - g(\boldsymbol{r})]F[g(\boldsymbol{r})] = F[f(\boldsymbol{r})] \,, \tag{3.45}$$

where $f(\boldsymbol{r})$ and $g(\boldsymbol{r})$ are arbitrary functions and $F[f(\boldsymbol{r})]$ is an arbitrary functional. It is obvious from this definition that the value of the delta functional $\delta[f(\boldsymbol{r})-g(\boldsymbol{r})]$ becomes infinite when two functions $f(\boldsymbol{r})$ and $g(\boldsymbol{r})$ are identical, and vanishes otherwise.

The Fourier transformation of the delta functional is defined in a similar way to the Fourier transformation of the delta function defined in (2.19):

$$\begin{aligned}
\delta[f(\boldsymbol{r})] &\equiv \lim_{L\to\infty} \delta\big(f(\boldsymbol{r}_1)\big)\ldots\delta\big(f(\boldsymbol{r}_L)\big) \\
&= \lim_{L\to\infty} \left(\frac{1}{2\pi}\right)^L \int \mathrm{d}g(\boldsymbol{r}_1)\ldots\int \mathrm{d}g(\boldsymbol{r}_L) \\
&\qquad \times \exp\Big\{\mathrm{i}\big[f(\boldsymbol{r}_1)g(\boldsymbol{r}_1) + \cdots + f(\boldsymbol{r}_L)g(\boldsymbol{r}_L)\big]\Big\} \\
&= \int \mathcal{D}\{g(\boldsymbol{r})\} \exp\left[\mathrm{i}\int f(\boldsymbol{r})g(\boldsymbol{r})\mathrm{d}\boldsymbol{r}\right] \,.
\end{aligned} \tag{3.46}$$

Substituting (3.44) into (3.43), we obtain

$$Z=\int\mathcal{D}\{\varphi(\boldsymbol{r})\}\int\mathcal{D}\{V(\boldsymbol{r})\}\exp\left\{-\beta\left[-\frac{1}{\beta}\ln\mathcal{Z}[V]+\mathcal{W}[\varphi]-\int V(\boldsymbol{r})\phi(\boldsymbol{r})\mathrm{d}\boldsymbol{r}\right]\right\}\,, \tag{3.47}$$

where we have introduced $V(\boldsymbol{r})\equiv w(\boldsymbol{r})/(\mathrm{i}\beta)$ and changed the integration path from $\int_{-\infty}^{\infty}\to\mathrm{i}\int_{-\mathrm{i}\infty}^{\mathrm{i}\infty}$ by changing the integration variable from $w(\boldsymbol{r})$ to $V(\boldsymbol{r})$, and we have neglected the factor $\mathrm{i}\beta$ that arises in the Jacobian because such a constant factor in the partition function does not contribute to the thermodynamic properties of the system. We have also introduced the quantity $\mathcal{Z}[V(\boldsymbol{r})]$ defined by

$$\begin{aligned}\mathcal{Z}[V(\boldsymbol{r})]&\equiv\frac{1}{M!}\int\mathrm{d}\Gamma\exp\left\{-\beta\left[\hat{\mathcal{H}}_0(\Gamma)+\int\hat{\phi}(\boldsymbol{r};\Gamma)V(\boldsymbol{r})\mathrm{d}\boldsymbol{r}\right]\right\}\\&=\frac{1}{M!}\left\{\int\mathrm{d}\Gamma_1\exp\left[-\frac{3}{2b^2}\sum_{i=0}^{N-1}|\boldsymbol{r}_i-\boldsymbol{r}_{i+1}|^2-\beta\sum_{i=0}^{N}V(\boldsymbol{r}_i)\right]\right\}^M\\&\equiv\frac{1}{M!}\mathcal{Z}_1^M\,,\end{aligned} \tag{3.48}$$

where $\mathrm{d}\Gamma_1=\mathrm{d}\boldsymbol{r}_0\ldots\mathrm{d}\boldsymbol{r}_N$ is the volume element of the phase space for a single chain. The first line is a general form and the second line is the particular expression for the polymer system. The definition (3.48) indicates that $\mathcal{Z}_1$ is the partition function of a single chain in an external potential $V(\boldsymbol{r})$. Hence, $\mathcal{Z}$ in (3.47) is the partition function of the ideal system composed of M independent chains described by the single chain partition function (3.48). We should also note that $\mathcal{Z}_1$ can be obtained by integrating the path integral defined by (3.3) with respect to the positions of both ends.

The effects of the thermal fluctuation in the mean field are neglected within the mean field approximation that was used when defining the path integral. In terms of the partition function (3.47), such a mean field approximation is realized by replacing the sum over the canonical ensemble with a Gaussian integral around the most probable state that minimizes the argument of the exponential function on the right-hand side of (3.47). Such an approximation is called the steepest descent method (see Fig. 3.21). The condition for determining the most probable state is given using the functional derivative defined in Sect. 3.3.1 as follows:

$$\left\{\begin{aligned}&\frac{\delta}{\delta\phi(\boldsymbol{r})}\big[\text{argument of the exponential function in (3.47)}\big]=0\,,\\&\frac{\delta}{\delta V(\boldsymbol{r})}\big[\text{argument of the exponential function in (3.47)}\big]=0\,.\end{aligned}\right. \tag{3.49}$$

Evaluating these functional derivatives, we obtain

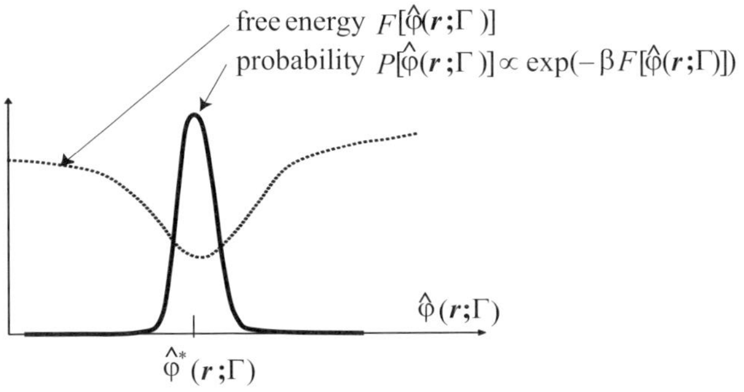

Fig. 3.21. The concept of steepest descent. The canonical probability distribution of finding the system with $\hat{\phi}(\boldsymbol{r};\Gamma)$ takes its maximum value at $\hat{\phi}^*(\boldsymbol{r};\Gamma)$ which minimizes the free energy functional $F[\hat{\phi}(\boldsymbol{r};\Gamma)]$. If the total number of degrees of freedom of the system is large enough, the canonical probability distribution peaks sharply around $\hat{\phi}^*(\boldsymbol{r};\Gamma)$. The canonical probability distribution can thus be well approximated by a Gaussian distribution centered at $\hat{\phi}^*(\boldsymbol{r};\Gamma)$

$$\begin{cases} V(\boldsymbol{r}) = \dfrac{\delta \mathcal{W}[\phi]}{\delta\phi(\boldsymbol{r})} , \\ \varphi(\boldsymbol{r}) = -\dfrac{1}{\beta\mathcal{Z}}\dfrac{\delta\mathcal{Z}}{\delta V(\boldsymbol{r})} = \langle\hat{\phi}(\boldsymbol{r};\Gamma)\rangle = \phi(\boldsymbol{r}) . \end{cases} \tag{3.50}$$

The derivation of the second equation is performed using the first line of (3.48). This result shows that the dummy function $\varphi(\boldsymbol{r})$ and the coarse-grained segment density $\phi(\boldsymbol{r})$ are the same within the mean-field approximation. Using this result, one can obtain the final expression for the free energy functional as

$$F[\phi(\boldsymbol{r})] = -k_{\mathrm{B}}T\ln\mathcal{Z}[V(\boldsymbol{r})] + \mathcal{W}[\phi(\boldsymbol{r})] - \int \mathrm{d}\boldsymbol{r} V(\boldsymbol{r})\phi(\boldsymbol{r}) . \tag{3.51}$$

The first term on the right-hand side of (3.51) is the free energy of ideal chains, each of which interacts independently with the self-consistent field. The second and third terms are the corrections for the double counting of the pair interactions among the segments and the elimination of the contribution from the artificial constraint force $\gamma(\boldsymbol{r})$ in (3.1).[2] We can also understand (3.51) as a Legendre transformation which transforms the free energy as a functional of the self-consistent field $V(\boldsymbol{r})$ (the first term) to a free energy as a functional of the segment density $\phi(\boldsymbol{r})$.

[2] For example, consider the pair interaction given by $\mathcal{W}[\hat{\phi}(\boldsymbol{r};\Gamma)] \propto (1/2)\int[\hat{\phi}(\boldsymbol{r};\Gamma)]^2\mathrm{d}\boldsymbol{r}$. Using the relations in (3.50), one can show that the second and third terms on the right-hand side of (3.51) give $-W[\phi(\boldsymbol{r})]$, which corrects for the doubly counted pair interactions.

The expression (3.51) is completely general except for the fact that $\mathcal{Z}$ contains information about the chain conformation. Therefore the characteristic properties of polymers enter into this free energy through $\mathcal{Z}$, calculated using the path integral as in (3.48).

Free Energy of Polymer Mixtures

Up to now, we have been discussing only one-component polymer solutions. It is easy to extend the argument to multicomponent polymer mixtures. The free energy for a homopolymer mixture is given by

$$F[\{\phi_K(\boldsymbol{r})\}] = -k_{\mathrm{B}}T \ln \mathcal{Z}[\{V_K(\boldsymbol{r})\}] + \mathcal{W}[\{\phi_K(\boldsymbol{r})\}] - \sum_K \int \mathrm{d}\boldsymbol{r} V_K(\boldsymbol{r})\phi_K(\boldsymbol{r}) \,, \tag{3.52}$$

where $K = A, B, \ldots,$ is the index that specifies the segment type. (The expression becomes more complex for systems containing block copolymers. We will not show the explanations of the free energy for such complex systems in this book.) The partition function $\mathcal{Z}$ of the system is given by

$$\mathcal{Z} = \prod_K \left[\frac{1}{M_K!} (\mathcal{Z}_{1K})^{M_K} \right] \,, \tag{3.53}$$

where M_K is the total number of K-type chains in the system and $\mathcal{Z}_{1K}$ is the single-chain partition function of a K-type polymer in the self-consistent field, whose explicit expression is obtained using (3.48) as

$$\mathcal{Z}_{1K} \equiv \int \mathrm{d}\Gamma_{1K} \exp\left[-\frac{3}{2b^2} \sum_{i=0}^{N_K-1} |\boldsymbol{r}_i - \boldsymbol{r}_{i+1}|^2 - \beta \sum_{i=0}^{N} V_K(\boldsymbol{r}_i) \right] \,, \tag{3.54}$$

where $\mathrm{d}\Gamma_{1K}$ is the volume element of the phase space for a single chain of K-type polymer and $V_K(\boldsymbol{r})$ is the self-consistent field potential acting on the K-type segment. We have also assumed that the effective bond length b is common to all segment species.

Free Energy of Homogeneous Mixtures

We now show that, for homogeneous polymer mixtures, the free energy (3.52) reduces to the Flory–Huggins free energy (2.67).

For a homogeneous mixture, both $\phi_K(\boldsymbol{r})$ and $V_K(\boldsymbol{r})$ are spatially uniform, and can be denoted by $\bar{\phi}_K$ and $\bar{V}_K$, both of which are constant. In such a case, the single-chain partition function $\mathcal{Z}_{1K}$ defined in (3.54) is rewritten as

$$\begin{aligned} \mathcal{Z}_{1K} &= \int \mathrm{d}\Gamma_{1K} \exp\left[-\frac{3}{2b^2} \sum_{i=0}^{N_K} |\boldsymbol{r}_i - \boldsymbol{r}_{i+1}|^2 - \beta(N_K+1)\bar{V}_K \right] \\ &= \exp\left[-\beta(N_K+1)\bar{V}_K \right] \mathcal{Z}_{0K} \,, \end{aligned} \tag{3.55}$$

where $\mathcal{Z}_{0K}$ is the partition function of a K-type ideal chain defined by (2.33). For this homogeneous system, the definition of the self-consistent field (3.50) reduces to

$$V_K(\boldsymbol{r}) = \frac{\delta \mathcal{W}[\{\phi_K(\boldsymbol{r})\}]}{\delta\phi(\boldsymbol{r})} = k_{\mathrm{B}}T \sum_{K'} \chi_{KK'}\phi_{K'}(\boldsymbol{r}) \longrightarrow k_{\mathrm{B}}T \sum_{K'} \chi_{KK'}\bar{\phi}_{K'} , \tag{3.56}$$

where we have assumed that the segment interaction $\mathcal{W}$ takes the form

$$\mathcal{W}[\{\phi_K(\boldsymbol{r})\}] = \frac{k_{\mathrm{B}}T}{2} \sum_{KK'} \chi_{KK'} \int \mathrm{d}\boldsymbol{r}\, \phi_K(\boldsymbol{r})\phi_{K'}(\boldsymbol{r}) . \tag{3.57}$$

The total number of K-type chains M_K and the average number density of K-type segments $\bar{\phi}_K$ are related by the equation

$$M_K = \frac{\mathcal{V}\bar{\phi}_K}{N_K + 1} , \tag{3.58}$$

where $\mathcal{V}$ is the system volume. Then, substituting (3.55), (3.56) and (3.58) into (3.52), we obtain the expression

$$\frac{F[\{\bar{\phi}_K\}]}{\mathcal{V}k_{\mathrm{B}}T} = \sum_K \frac{\bar{\phi}_K}{N_K} \ln \bar{\phi}_K + \frac{1}{2} \sum_{KK'} \chi_{KK'}\bar{\phi}_K\bar{\phi}_{K'}$$
$$+0\,\text{th and } 1\,\text{st order terms in } \bar{\phi}_K , \tag{3.59}$$

where we have assumed $N_K \gg 1$. (For a solvent with the same size as the segment, $N_K = 1$.) Equation (3.59) is the same as the Flory–Huggins free energy presented in (2.67), where the volume fraction is replaced by the number density.

Phase Equilibrium of Multicomponent Systems and Chemical Potential

In order to calculate the phase diagram of a multicomponent system using the self-consistent field theory, we have to evaluate the chemical potential of the two coexisting phases separated by an interface. In such an inhomogeneous multicomponent system, the chemical potential of each component is not constant, even in the equilibrium state. Let us denote the chemical potential of the K-type component at $\boldsymbol{r}$ by $\mu_K(\boldsymbol{r})$. By definition, $\mu_K(\boldsymbol{r})$ is the energy required to add a single molecule of the K-type component at the position $\boldsymbol{r}$ from outside of this system.

The chemical potential μ of a uniform equilibrium system is given by the relation $\mu = \partial F/\partial N$, where F and N are the total free energy of the system and the total number of molecules contained in the system, respectively. The chemical potential of an inhomogeneous system can be defined in a similar manner. When the free energy of the system is given as a functional $F[\{\phi_K(\boldsymbol{r})\}]$ of the density distributions $\{\phi_K(\boldsymbol{r})\}$, the chemical potential of this system is given using the functional derivative as

$$\mu_K(\boldsymbol{r}) = \frac{\delta F}{\delta \phi_K(\boldsymbol{r})} . \tag{3.60}$$

Here, we have used the so-called local equilibrium assumption, where the inhomogeneous system is divided into many small cells each of which can be regarded as a uniform equilibrium system. (Recall the lattice description used in Sect. 3.3.1.) If an exchange of molecules between the system and the environment is prohibited (i.e., we have a closed system), a necessary condition for the system to be in equilibrium is the vanishing of the mutual diffusion between any pair of species K and K'. Thus, $\mu_K(\boldsymbol{r}) - \mu_{K'}(\boldsymbol{r})$ becomes everywhere constant for arbitrary pairs of components K and K', i.e., the chemical potential difference is constant, although the segment density varies from place to place.

We now derive an explicit expression for the chemical potential (3.60) using an intuitive argument. In order to perform the functional derivative in (3.60), we need an explicit expression for the free energy F as a functional of the segment density distribution $\{\phi_K^{\mathrm{target}}(\boldsymbol{r})\}$ that is in general different from its equilibrium distribution. To do this, the free energy (3.51) should be supplemented by the constraint condition that the segment density $\{\phi_K(\boldsymbol{r})\}$ should be equal to the specified segment density $\{\phi_K^{\mathrm{target}}(\boldsymbol{r})\}$. We introduce a Lagrange multiplier $\{\gamma_K(\boldsymbol{r})\}$ for this constraint and add an extra term to the free energy, viz.,

$$F + \sum_K \int \mathrm{d}\boldsymbol{r} \gamma_K(\boldsymbol{r}) \left[\phi_K(\boldsymbol{r}) - \phi_K^{\mathrm{target}}(\boldsymbol{r})\right] . \tag{3.61}$$

Here, the Lagrange multiplier $\{\gamma_K(\boldsymbol{r})\}$ is the constraint force that fixes the segment density to $\{\phi_K^{\mathrm{target}}(\boldsymbol{r})\}$. When the system described by (3.61) is equilibrated, the thermodynamic force acting on each segment, i.e., the chemical potential $\mu_K(\boldsymbol{r})$, is exactly cancelled by the constraint force $\{\gamma_K(\boldsymbol{r})\}$. Therefore, we have the relation

$$\mu_K(\boldsymbol{r}) = -\gamma_K(\boldsymbol{r}) . \tag{3.62}$$

In the actual numerical calculation, the multiplier $\gamma_K(\boldsymbol{r})$ is obtained by an iteration procedure so that the segment density field calculated using the path integral will coincide with the specified profile $\phi_K^{\mathrm{target}}(\boldsymbol{r})$.

3.3.3 Numerical Solutions of Self-Consistent Field Equations

In Sect. 3.2.3, we discussed how to solve the self-consistent field equations analytically using the classical approximation. Although such an analytic method is useful for understanding the qualitative features of the system, it is not usually enough to make quantitative predictions of real materials due to the poor accuracy of the approximation. For quantitative predictions, numerical analysis of the self-consistent field equations is widely and successfully used.

Figures 3.23 and 3.25 show typical examples of phenomena for which the self-consistent field theory has been applied. (Some of these examples will be discussed in later sections.) Figure 3.23 shows disordered structures without any periodicity while Fig. 3.25 shows regular structures with crystalline symmetries. For these two types of system, different numerical methods are used. In the following, we discuss several recent studies using such numerical methods.

General Procedure

In order to calculate mesoscopic structures in polymer systems using the self-consistent field theory, we have to solve the following set of equations:

(a) definition (3.1) of the self-consistent field,
(b) statistical weight (3.10) of the conformation of an ideal chain in the self-consistent field (path integral),
(c) expression (3.11) for the segment density in terms of the path integral,
(d) expression (3.52) for the free energy.

Items (a)–(c) include three unknown functions, i.e., the segment density $\phi(\boldsymbol{r})$, the self-consistent field $V(\boldsymbol{r})$, and the path integral $Q(0, \boldsymbol{r}_0; s, \boldsymbol{r})$. These three functions should be determined in such a way that they do not contradict each other. This is the self-consistency condition. Once (a)–(c) have been solved self-consistently, the free energy can be evaluated by substituting the results into (d), and the stability of the structure is then evaluated. Finally, the phase diagram is obtained by evaluating the free energy for all candidate for the equilibrium structure.

In order to solve the self-consistent set of equations (a)–(c) numerically, an iteration method is often adopted. A simple illustration of the iteration method is given below.

Let us consider a self-consistent equation

$$x = f(x) \, . \tag{3.63}$$

In the iteration method, the following procedure is repeated until the variable x_i converges to a certain value:

$$\begin{cases} x_0 \leftarrow \text{initial guess} \, , \\ x_1 = f(x_0) \, , \\ x_2 = f(x_1) \, , \\ \quad \vdots \, . \end{cases} \tag{3.64}$$

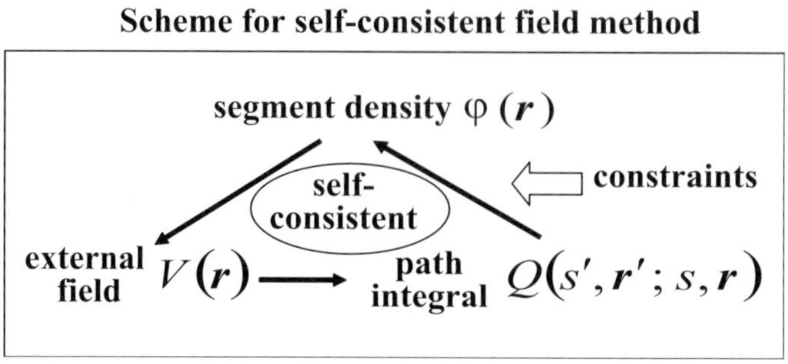

Fig. 3.22. Concept of the iterative procedure for the self-consistent field theory using the path integral

The same procedure is applicable to the self-consistent field theory of polymers. We set appropriate initial values for $\{\phi_K(\boldsymbol{r})\}$ and $\{V_K(\boldsymbol{r})\}$, and repeatedly update these variables using (a)–(c) until the fields converge. When updating the variables, the external constraint conditions (e.g., the incompressibility condition) should be taken into account. The details of this updating method can be found in the references [26]. In Fig. 3.22, we show the concept of the iterative procedure for the self-consistent field theory.

There are two types of iterative scheme for the self-consistent field equations, depending on whether the system has spatial symmetry or not:

- calculations in the real space,
- calculations in the reciprocal lattice space (Fourier space).

Instead of going into the details of these methods, we will discuss only the concepts underlying them.

Real-Space Calculations

When no spatially periodic structures are expected in the target system, the calculations should be done in the real space [26]. Typical examples of such systems are:

(a) Irregular domain structures generated by phase separation after quenching a uniform mixture of polymers or block copolymers.
(b) Interfaces between domains.
(c) Systems containing a special point or points that destroy the translational symmetry of the system, such as polymer brushes grafted onto a solid surface.
(d) Polymer solutions where colloidal particles or solid microparticles are irregularly dispersed. (Such particles are called fillers.)
(e) Dilute solutions of molecular aggregates.

These examples are shown schematically in Fig. 3.23.

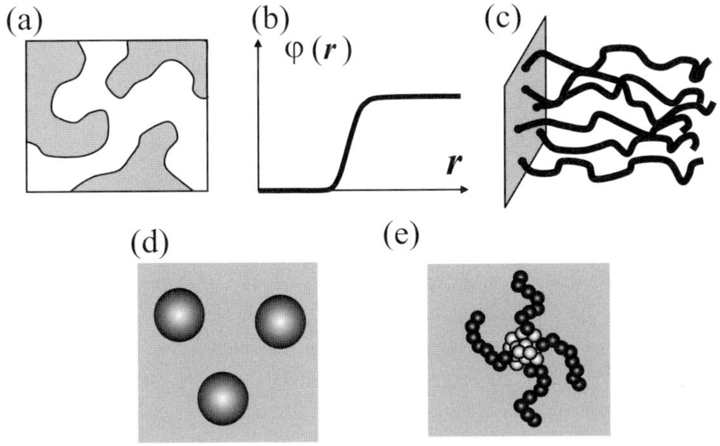

Fig. 3.23. Examples of periodic structures. (**a**) Domain structures of phase separation, (**b**) interfaces, (**c**) polymer brushes, (**d**) colloidal particles, and (**e**) molecular aggregates (micelles)

The irregular domains shown in Fig. 3.23a are often obtained by quenching a uniformly mixed polymer mixture. In this case, the final equilibrium structure may be an ordered structure with regularly arranged domains. However, the domain structures during the relaxation process toward the final equilibrium state are usually irregular due to the existence of many defects. The translational symmetry of the system is also broken when it contains singularities, such as interfaces (Fig. 3.23b), solid surfaces (Fig. 3.23c) and impurities (Fig. 3.23d).

Molecular aggregates (Fig. 3.23e) are often formed in a dilute solution of amphiphilic molecules. The amphiphilic molecule is a molecule composed of two chemically different parts, each of which dissolves in different solvents. A block copolymer is a typical example of an amphiphilic molecule. Aggregates formed by amphiphilic molecules, such as the one shown in Fig. 3.24, are called micelles. The lowest density of an amphiphilic solution at which micelles can be observed is called the critical micelle concentration (CMC). This CMC is extremely small for solutions of block copolymers with large molecular weights. To predict the CMC using the self-consistent field theory, the real-space calculation of a system containing a single micelle is often performed because the average separation between the micelles at CMC is so large that the micelles can be regarded as mutually independent [26].

In the real-space calculation, we adopt a discretization of the space by introducing a spatial lattice (spatial mesh) structure, and describe the segment density $\phi_K(\boldsymbol{r})$, the self-consistent field $V_K(\boldsymbol{r})$, and the path integral $Q_K(0, \boldsymbol{r}_0; s, \boldsymbol{r})$ as sets of values at the lattice points (mesh points). For example, by discretizing x, y and z coordinates with the same mesh width Δ, we

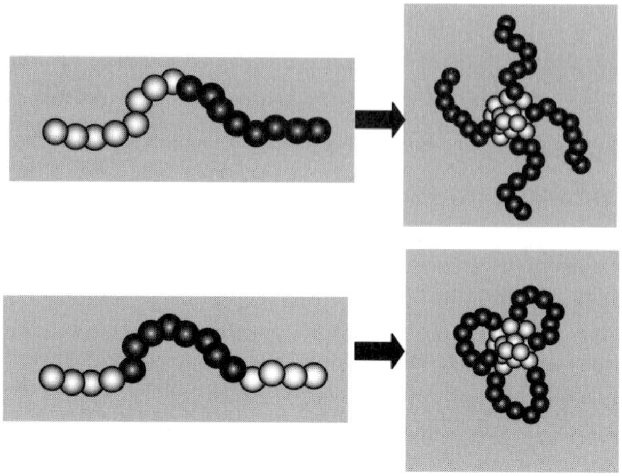

Fig. 3.24. Schematic pictures of the micelles formed by block copolymer chains in a solvent. (**a**) The case with a diblock copolymer and (**b**) the case with a triblock copolymer. In these pictures, the *white segments* of the block copolymers have low affinity to the solvent so that these segments coagulate to form the core of the micelles

can generate a set of lattice points $(l\Delta, m\Delta, n\Delta)$ $(l, m, n = 0, \pm 1, \pm 2, \ldots)$. Then the field variables are

$$\phi_K(\boldsymbol{r}) \longrightarrow \{\phi_K(l\Delta, m\Delta, n\Delta)\} \equiv \{\phi_K(l, m, n)\} \; . \tag{3.65}$$

At the same time, the evolution equation for the path integral (3.10) is rewritten as a finite difference equation, which is solved numerically. The discretized version of the Laplacian of any function $f(\boldsymbol{r})$ is obtained by approximating $\nabla^2 f(\boldsymbol{r})$ at the lattice point (l, m, n) by a linear combination of the value of the function $f(\boldsymbol{r})$ at the point $f(l, m, n)$, that of the nearest neighbor lattice points $\sum_{nn} f(l, m, n)$ and that of the next nearest neighbor lattice points $\sum_{nnn} f(l, m, n)$. Then $\nabla^2 f(\boldsymbol{r})$ is discretized as

$$\nabla^2 f(\boldsymbol{r}) \longrightarrow C_0 f(l, m, n) + C_1 \sum_{nn} f(l, m, n) + C_2 \sum_{nnn} f(l, m, n) \; , \tag{3.66}$$

where the coefficients C_0, C_1 and C_2 are determined so that the leading term in the Taylor series expansion of the right-hand side around the lattice point (l, m, n) coincides with the left-hand side. There is still an infinite number of choices of these coefficients C_0, C_1 and C_2 under this constraint. The coefficients are usually chosen so that they guarantee the isotropy of the Laplacian operator. One of the commonly used sets of coefficients for 3-dimensional lattices is $C_0 = -9/(2\Delta^2)$, $C_1 = 1/(2\Delta^2)$ and $C_2 = 1/(8\Delta^2)$.

Once we have discretized the variables and the equations, we have only to repeat the iteration procedure as in (3.64) for the self-consistent field, the segment density, and the path integral.

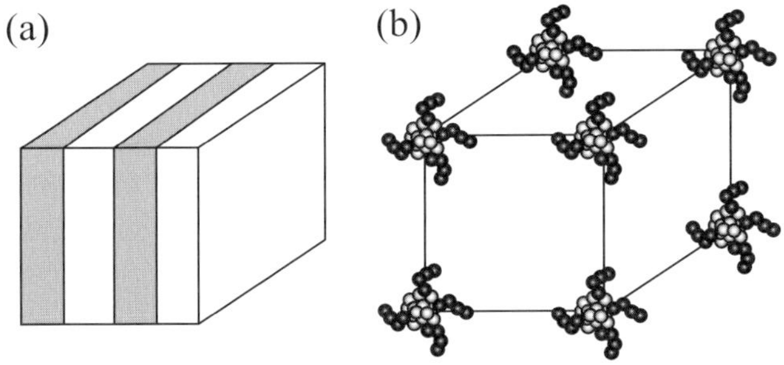

Fig. 3.25. Examples of periodic structures. (**a**) Microphase-separated domain structures and (**b**) regular arrangements of molecular aggregates

Reciprocal Lattice Space (Fourier Space) Calculations

This method is applicable when the target system has a crystalline symmetry [27]. Typical examples are:

- equilibrium domain structures of the microphase separation of block copolymer melts,
- regularly arranged dense molecular aggregates.

Schematic illustrations of these systems are shown in Fig. 3.25.

The block copolymer melt (Fig. 3.25a) exhibits various microphase separated structures where layered, cylindrical or globular domains are regularly arranged. When a concentration of a micellar solution is increased (Fig. 3.25b), the micelles often show a crystal structure due to the repulsive interaction between them. For these periodic structures, representations and analyses in the reciprocal lattice space (or Fourier space) are useful.

As we discussed in Sect. 2.3.2, the square of the absolute value of the Fourier transform of the segment density distribution, i.e., $|\delta\tilde{\phi}(\boldsymbol{q})|^2$, corresponds to the scattering function [see (2.46) and (2.48)]. Note that the discussion in Sect. 2.3.2 was for scattering from a single isolated chain. The same procedure using the Fourier transform also applies to the many-chain systems we are discussing here. The scattering function of a system with a crystalline symmetry is composed of two contributions: one is the scattering from the frame of the crystal structure, and the other is the scattering from the unit structure placed on each of the lattice points, such as micelles and phase-separated domains. The former is called the structure factor, while the latter is called the form factor. The scattering function from a micellar crystal is therefore a superposition of the structure factor of the crystalline structure and the form factor of a single isolated micelle (see Fig. 3.26).

The actual procedure of the self-consistent field calculation in the reciprocal lattice space is as follows. We use the Fourier transform of (3.10) with respect to the segment index s and the position $\boldsymbol{r}$. If there is no self-consistent

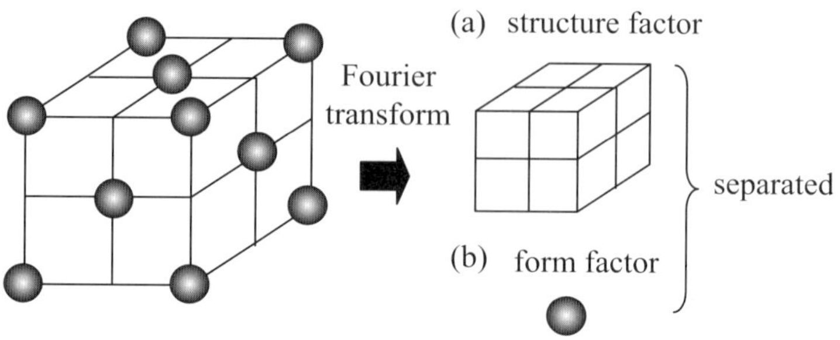

Fig. 3.26. Structure and scattering function of a system with crystalline symmetry. The scattering function is a product of (**a**) the structure factor and (**b**) the form factor

field, the Fourier transform of (3.10) with respect to s leads to the eigenvalue equation

$$\nabla^2 f_n(\boldsymbol{r}) = -\lambda_n D^{-2} f_n(\boldsymbol{r}) \;, \tag{3.67}$$

where $\{f_n(\boldsymbol{r})\}$ are orthonormal eigenfunctions and the λ_n are the corresponding eigenvalues. In (3.67), we explicitly expressed the dependence on the periodicity D of the structure. The solution of (3.10) with the self-consistent field can be expressed as a linear combination of these eigenfunctions $\{f_n(\boldsymbol{r})\}$.

To take the crystalline symmetry into account, we expand the eigenfunction $f_n(\boldsymbol{r})$ in terms of the reciprocal lattice vectors:

$$f_n(\boldsymbol{r}) = \frac{1}{\sqrt{N_n}} \sum_{j\in n} S_j^n \exp\left(\mathrm{i}\boldsymbol{G}_j^n \cdot \boldsymbol{r}\right) \;, \tag{3.68}$$

where n is the index of the eigenfunction, N_n is the total number of reciprocal lattice vectors of the nth eigenfunction, $\boldsymbol{G}_j^n$ is the jth reciprocal lattice vector of the nth eigenfunction, and S_j^n is ± 1. When the crystal symmetry is specified, the values of N_n, $\{\boldsymbol{G}_j^n\}$ and $\{S_j^n\}$ are uniquely determined. Therefore, expanding the path integral $Q(0, \boldsymbol{r}_0; s, \boldsymbol{r})$ in terms of the orthonormal basis set $\{f_n(\boldsymbol{r})\}$ and substituting (3.68), we can obtain a Fourier transform of the path integral with respect to both s and $\boldsymbol{r}$. Using the same Fourier transformation, the self-consistent set of equations (3.10), (3.11) and (3.52) are transformed into a linear matrix equation for the Fourier amplitudes, which can be solved using a standard numerical method for linear algebraic equations [28].

The procedure to obtain the phase diagram is as follows. First, one has to choose an appropriate crystal symmetry and evaluate the free energy as a function of the period D, i.e., $F(D)$, using the above-mentioned procedure. The most stable period D^* within the specified crystal symmetry is obtained as the one that minimizes $F(D)$, i.e.,

$$F(D^*) = \min_D F(D) \;. \tag{3.69}$$

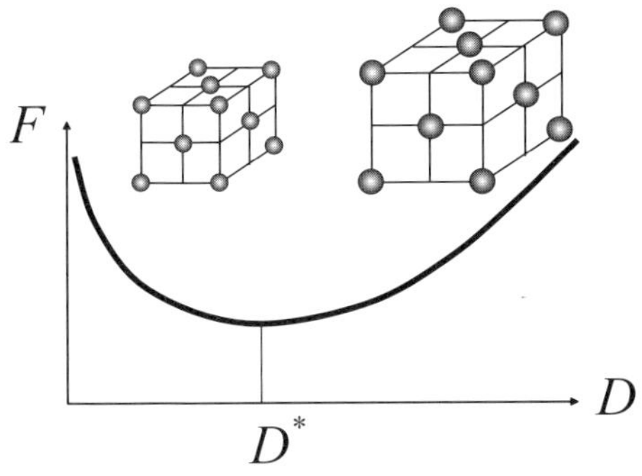

Fig. 3.27. In order to find the equilibrium state, the minimum value of the free energy within the assumed symmetry is calculated (shown by D^* in the figure). Then the minimum values of the free energy for different crystal symmetries are compared

This procedure is illustrated schematically in Fig. 3.27. Repeating this procedure for all possible crystal symmetries, we can find the equilibrium structure that gives the minimum free energy $F(D^*)$ among all the candidates for the crystal symmetries of the equilibrium structure.

Dynamical Simulations Based on Self-Consistent Field Theory

To obtain the equilibrium structure using the self-consistent field theory, we have to look for a state where the chemical potential difference between an arbitrary pair of the segment types is constant. One of the real-space simulation methods for this purpose is the dynamic simulation method where the segment density $\phi_K(\boldsymbol{r})$ is updated according to a dynamical process until the system reaches its final equilibrium state. (To accelerate the convergence of the scheme, we often use unrealistic dynamics that does not correspond to the dynamics of real systems.)

A realistic dynamical simulation model is based on the diffusion dynamics of the segments [29–31]. In this model, we combine the conservation law for the segment density and the linear diffusion law. The conservation law is expressed by the continuity equation

$$\frac{\partial}{\partial t}\phi_K(\boldsymbol{r},t) + \nabla\cdot\boldsymbol{j}_K(\boldsymbol{r},t) = 0 \, . \tag{3.70}$$

The linear diffusion process is characterized by Fick's law

$$\boldsymbol{j}_K(\boldsymbol{r},t) = -\sum_{K'} L_{KK'}\nabla\mu_{K'}(\boldsymbol{r},t) \, , \tag{3.71}$$

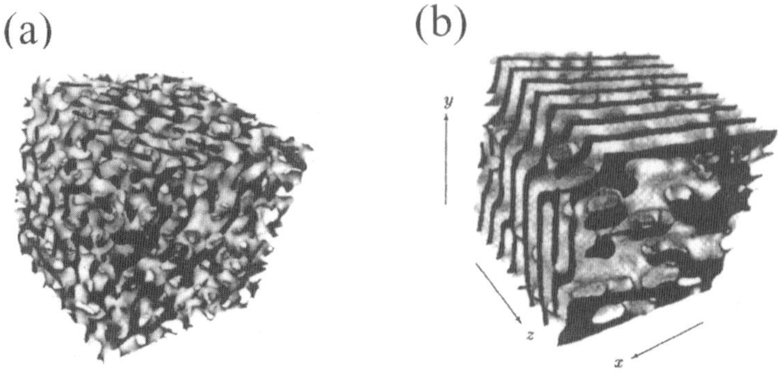

Fig. 3.28. Structures of an AB block copolymer melt (**a**) before and (**b**) after the shear is imposed. Reprinted with permission from A.V. Zvelindovsky et al., Phys. Rev. E **57**, R4879 (1998). (© 1998, American Physical Society)

where $L_{KK'}$ is the kinetic coefficient measuring the mobility of the segments. Fick's law says that the diffusion current of the segment is proportional to the spatial gradient of the chemical potential.

These two equations coupled with the expression for the chemical potential (3.62) determine the time evolution of the segment density fields, which can be simulated using numerical techniques.

Using the dynamical self-consistent field theory introduced here, one can simulate various dynamics in polymer systems. Figure 3.28 is the simulation result for a shared AB diblock copolymer melt. In the initial state (Fig. 3.28a), the system shows an irregular domain structure. When we impose a shear flow on this system, a well-aligned lamellar microstructure is formed (Fig. 3.28b). This simulation shows that an externally imposed flow field has a crucial effect on the ordering of the domains.

3.3.4 Examples of Numerical Simulations Using Self-Consistent Field Theory

In this section, we show several examples of numerical simulations using the self-consistent field theory.

Predicting Microphase-Separated Structures of Block Copolymer Systems

As discussed in Fig 3.2 of Sect. 3.1, block copolymer melts undergo microphase separation when the temperature becomes lower than a certain critical temperature. From the point of view of material engineering, an important feature of such microphase-separated structures is the uniformity

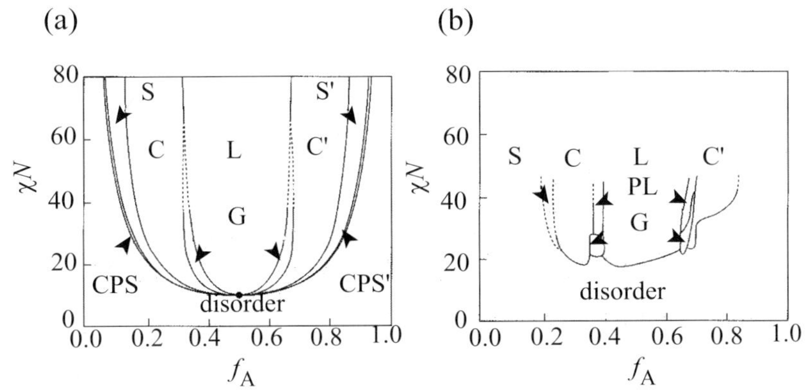

Fig. 3.29. Phase diagrams for an AB-type block copolymer melt obtained by (**a**) numerical self-consistent field calculation and (**b**) experiment on a polystyrene–polyisoprene block copolymer melt. (The latter is the same as Fig. 3.3.) The *ordinate* is χN and the *abscissa* is the block ratio defined by the ratio of the length of the A-subchain to the total chain length, i.e., $f = N_A/(N_A + N_B)$. The *symbols* for individual phases are L (lamellar), C (cylindrical), S (spherical), G (gyroid), PL (perforated lamellar) and CPS (close-packed spherical) phases. Reprinted with permission from F.S. Bates and G.H. Fredrickson, *Physics Today* **52**, 32 (1999). (© 1999, American Institute of Physics)

on the coarse-grained scale ($\geq$ 100 nm). Due to this uniformity, two or more polymers, each with special functions, can be combined to form a macroscopically uniform material. Thus, these microphase-separated structures are candidates for highly functional devices such as nanoscale switching devices and electron conducting devices. In designing these materials, the self-consistent field theory can be a useful tool.

Figure 3.29a shows the phase diagram of the microphase separation of an AB-type block copolymer obtained by numerical self-consistent field calculation with the Fourier representation. Figure 3.29b is the corresponding experimental phase diagram. (This phase diagram has already been presented in Fig. 3.3.) We confirm that the self-consistent field theory reproduces the overall behavior of the phase diagram semi-quantitatively. We should note here that the equilibrium microphase separated domain structure of a melt of a sufficiently long diblock copolymer is uniquely determined when the values of χN and f are specified, where χ is the interaction parameter between the two types of segment that compose the block copolymer, N is the total length of the block copolymer chain and f is the so-called block ratio defined as the ratio of the length of one of the blocks to the total chain length (see Sect. 3.1 and Exercise 3.1.) As discussed in the box entitled χ-*Parameters* on p. 61, the χ-parameter is inversely proportional to the temperature.

(a) 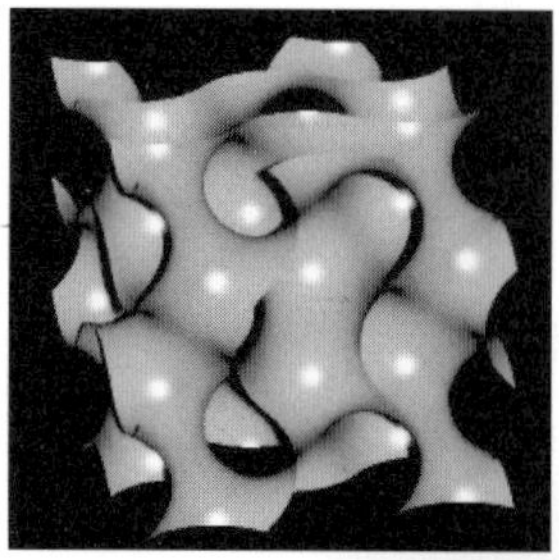
(b) 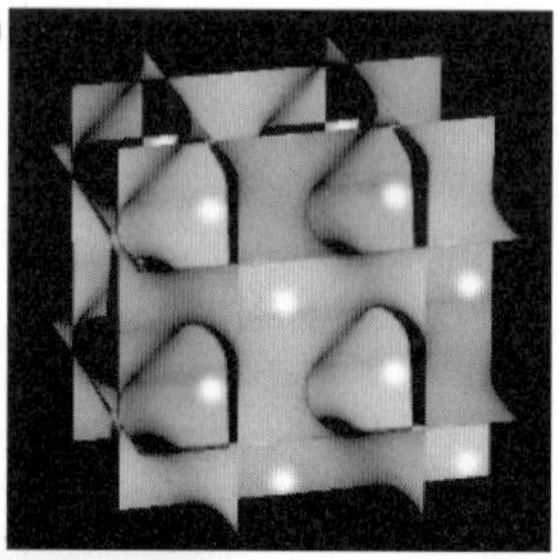
(c) 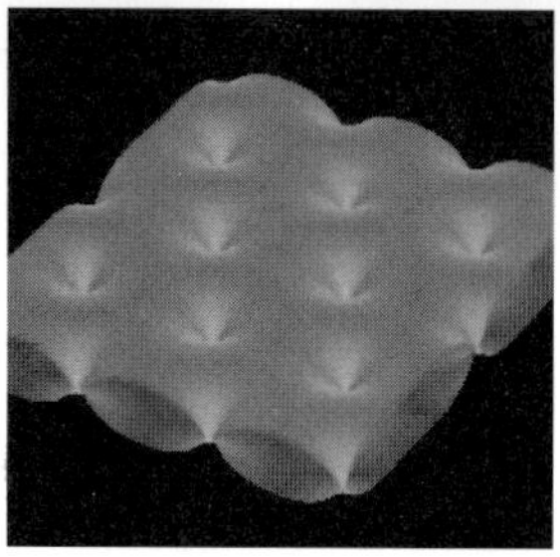

Fig. 3.30. Candidates for complex domain structures in microphase separations of *AB*-type block copolymers. (**a**) Gyroid, (**b**) ordered bicontinuous double diamond, and (**c**) perforated lamellar structures. Computer images produced by James T. Hoffman at the Mathematical Sciences Research Institute, Berkeley, U.S.A., with support from the Department of Energy. Original and related images and research can be found on the scientific graphics pages at www.msri.org

The ordinate of Fig. 3.29 can thus be regarded as the inverse temperature $1/T$.[3]

Apart from the well-known layered (lamellar), cylindrical, and spherical domain structures, there are much more complex domain structures such as the gyroid, ordered bicontinuous double diamond (OBDD) and perforated lamellar structures, which are also candidates for equilibrium domain structures (see Fig. 3.30). [The values of the parameters for these complex structures in (3.68), i.e., N_n, $\{\boldsymbol{G}_j^n\}$ and $\{S_j^n\}$, are known in crystallography and are published as a table.]

[3] Ordered phases, such as microphase-separated phases, usually appear in the lower temperature region. Such a phase diagram is said to be of the upper critical solution temperature (UCST) type. On the other hand, if the ordered phases appear in the high temperature region, the phase diagram is said to be of the lower critical solution temperature (LCST) type. The LCST phase diagram is realized when the mixing entropy effect in the high temperature phase is cancelled by other enthalpic contributions, such as hydrophobic interactions.

Strong Segregation Theory for Block Copolymers

As will be discussed later, a diblock copolymer melt with the block ratio $f = 0.5$ shows a critical point at $\chi N \cong 10.5 \equiv (\chi N)_c$ (see Sect. 4.1.3). If the value of χN is much larger than $(\chi N)_c$, the repulsive interaction between A and B segments is so large that the interface between A-phase and B-phase becomes sharp to avoid unfavorable contacts between AB segment pairs. Such a regime is called the strong segregation regime. In such a regime, the size and morphology of the microphase-separated domain structures are determined by competition between three major factors:

- Interfacial energy originating from unfavorable contact between different types of segment at the interface. (The excess free energy stored at the interface per unit interfacial area is called the interfacial tension.)
- Elastic energy due to stretching of the chain.
- Bending elastic energy of the interface, related to the interfacial curvature.

For simplicity, we first consider a lamellar structure of an AB diblock copolymer melt. As the interface in a lamellar structure is flat, the contribution from the interfacial bending elasticity is absent. Thus the equilibrium spacing of the lamellar layers is determined by competition between the interfacial energy and the elastic energy due to stretching of the chains.

Denoting the period of the lamellar layers by D, let us evaluate the equilibrium lamellar period D^* in the strong segregation limit. As the total area of the A/B interfaces is proportional to the number of interfaces per unit length in the direction perpendicular to the lamellar layers, the interfacial energy $F_{\text{interface}}$ is proportional to $1/D$ and is almost independent of the chain length N. (For the interfaces in polymer blends, the interfacial tension tends to a finite limiting value in the very-long-chain limit. The correction to this limiting value vanishes asymptotically as $1/N$ in the long-chain limit [32].) On the other hand, the elastic energy associated with chain stretching is evaluated using a Gaussian chain model. Equation (2.37) suggests that the elastic energy of a single chain due to stretching is proportional to D^2/N, where we used the lamellar period D as a measure of the chain extension. As the total number of chains in the system is proportional to $1/N$ due to the incompressibility condition, the total elastic energy $F_{\text{elasticity}}$ in the system due to chain stretching is proportional to $(D/N)^2$. Combining the interfacial contribution and the elastic contribution, the total free energy F of the system is given by

$$F = F_{\text{interface}} + F_{\text{elasticity}} = \frac{C_{\text{interface}}}{D} + C_{\text{elasticity}} \left(\frac{D}{N} \right)^2 , \tag{3.72}$$

where $C_{\text{interface}}$ and $C_{\text{elasticity}}$ are constant parameters. Minimizing this free energy with respect to the period D, we obtain the following power law for the equilibrium lamellar period D^* :

$$D^* \sim N^{2/3} . \tag{3.73}$$

This 2/3 power law dependence of the lamellar spacing on the chain length is confirmed by both experiment and numerical calculation using the self-consistent field theory [33, 34].

Curvature and Bending Elasticity of a Curved Surface

A point on a two-dimensional smooth curved surface in three-dimensional space is specified by a pair of parameters $\boldsymbol{a} \equiv (a_1, a_2)$. At the point $\boldsymbol{a}$, we cut this curved surface by two planes that cross orthogonally along the normal line of the surface. These two planes define two cutting curves, whose radii of curvature are denoted by R_1 and R_2, respectively. Rotating these two lines around the point $\boldsymbol{a}$, one can find the configuration where R_1 and R_2 become the maximum and the minimum. Let us denote these by R_{max} and R_{min}. The corresponding curvatures $1/R_{\mathrm{max}}$ and $1/R_{\mathrm{min}}$ are called the principal curvatures. The mean curvature C_{m} and the Gaussian curvature C_{g} are defined in terms of these principal curvatures by

$$C_{\mathrm{m}} \equiv \frac{1}{2}\left(\frac{1}{R_{\mathrm{min}}} + \frac{1}{R_{\mathrm{max}}}\right) \quad \text{and} \quad C_{\mathrm{g}} \equiv \frac{1}{R_{\mathrm{min}} R_{\mathrm{max}}} .$$

They are independent of the particular choice of coordinate system $\boldsymbol{a}$.

These curvatures are important for evaluating the bending elastic energy of surfaces that have internal structures, such as an interface onto which amphiphilic molecules are adsorbed or a membrane formed by block copolymer molecules. Due to the internal structure of the surface, bending of the surface causes an increase in the bending elastic energy. If the radius of curvature of the surface is much larger than the thickness of the surface (i.e., in the small curvature limit), the bending elastic energy can be approximated by keeping terms up to the second order in its Taylor series expansion with respect to the mean and Gaussian curvatures. This procedure leads to Helfrich's bending energy model:

$$H_{\text{bending elasticity}} = \int \mathrm{d}\boldsymbol{a} \Big\{ 2\kappa \big[C_{\mathrm{m}}(\boldsymbol{a}) - C_0 \big]^2 + \bar{\kappa} C_{\mathrm{g}}(\boldsymbol{a}) \Big\} , \tag{3.74}$$

where the integral $\int \mathrm{d}\boldsymbol{a}$ is taken over the whole surface. The parameter C_0 is called the spontaneous curvature. It gives a measure of how much the surface tends to bend when no external constraint is imposed. Helfrich's model is a useful starting point for studying the equilibrium shapes of interfaces and membranes.

If there is no breaking and no adhesion of the membrane, its topology is preserved even if the membrane deforms. In such a membrane with a constant topology, it is known that the Gaussian curvature term in (3.74) contributes a constant term to $H_{\text{bending elasticity}}$ due to the Gauss–Bonnet theorem [35]. In this case, only the mean curvature term in (3.74) need be considered. A typical example of this simple case is the red blood cell [36].

Distribution of Interfacial Curvature

When an interface is curved, an extra contribution to the free energy arises from the bending elasticity of the interface. As described in the box entitled *Curvature and Bending Elasticity of a Curved Surface*, there are two independent curvatures of a curved surface. One is the mean curvature and the other is the Gaussian curvature, both of which are independent of the choice of coordinate system. The shape of the curved surface is fully determined by these two curvatures, and the bending elastic energy is therefore also described by these two curvatures. If the radius of curvature R is much larger than the thickness of the interface, the bending free energy can be approximated by the leading few terms of its Taylor series expansion with respect to the mean and Gaussian curvatures. We usually truncate the expansion after the second order in $1/R$. Such a model is called Helfrich's model of bending elasticity [see (3.74)].

The well known Gauss–Bonnet Theorem tells us that the integral of the Gaussian curvature term in (3.74) over the whole interface gives a constant as long as the topology of the interface is preserved [35]. Therefore, if there is no topological change, the equilibrium shape of the interface is determined by the mean curvature alone. We should note here that vanishing mean curvature is not equivalent to a flat surface. The mean curvature vanishes when the two principal curvatures have the same magnitude but opposite signs, i.e., for a saddle-shaped surface. (The definition of the principal curvatures is given in the box entitled *Curvature and Bending Elasticity of a Curved Surface*.) For this reason, a complex interconnected domain structure where the interface takes the saddle-shape everywhere can be a stable structure of the bending elastic energy. Indeed, the gyroid and OBDD structures shown in Fig. 3.30 are not flat but have very small (or almost constant) mean curvature at every point. Such a surface with vanishing (or constant) mean curvature is called the minimal surface (or constant mean curvature surface).

Figure 3.31 illustrates the curvature distributions of the interfaces in microphase-separated block copolymer melts, calculated using the self-consistent field method. One can confirm that, except for the cylindrical structure, the curvature is not strictly constant. This reflects the fact that the equilibrium structure of a block copolymer melt is determined not only by the bending elasticity of the interface but also by the stretching energy of the chains that should fill the space between the interfaces.

Structural Phase Transitions in Block Copolymer Systems

The self-consistent field theory is useful not only for calculating the equilibrium structure but also for simulating a transition from one stable structure to another stable structure induced by a change in a thermodynamic variables such as the temperature. In such a transition, it is important to predict the dynamical pathway from the starting structure to the final structure.

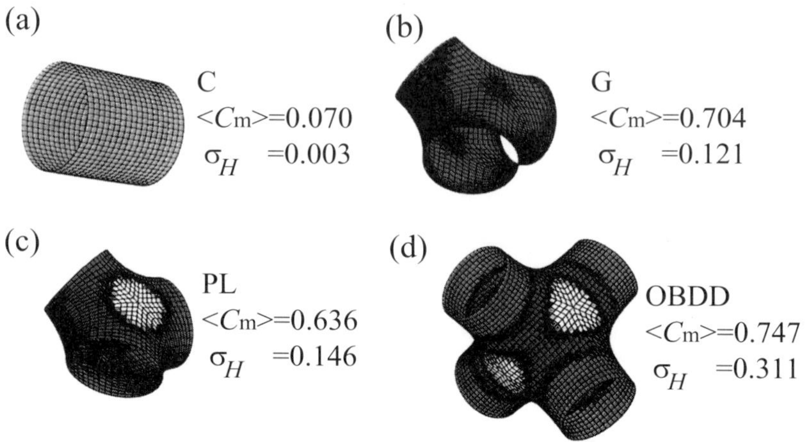

Fig. 3.31. Curvature distributions of the interfaces of the microphase separation of *AB*-type block copolymer melts. (**a**) Cylinder, (**b**) gyroid, (**c**) perforated lamellar, and (**d**) ordered bicontinuous double diamond structures, respectively. *White regions* have mean curvature close to 0 and *black regions* have large absolute values of the mean curvature. $\left\langle C_{\mathrm{m}} \right\rangle$ is the average of the mean curvature over the whole interface, and σ_{H} is its standard deviation. Reprinted with permission from M.W. Matsen and F.S. Bates, Macromolecules **29**, 7641 (1996). (© 1996, American Chemical Society)

For example, Fig. 3.32 shows the results of self-consistent calculations on the transition from the cylindrical structure to the spherical structure or the lamellar structure. In these calculations, the initial cylindrical structure is thermodynamically unstable. Even for such an unstable structure, the free energy can be evaluated using the self-consistent field calculation. A small perturbation is then added to the interfacial structure, with which numerical linear perturbation analyses are performed on the free energy around the cylindrical structure. This procedure leads to a linear algebraic equation with a coefficient matrix whose elements depend on the free energy of the starting cylindrical structure. The eigenvalue and eigenvector of this coefficient matrix give us the unstable modes with which the unstable cylindrical structure is destroyed.

In these figures, the most unstable mode is shown with changing amplitude. Thus, the figures do not correspond to the real dynamics. As the calculation is based on a linear analysis, the analysis may be invalid for the large deformations shown in the last figures of Figs. 3.32a and b. In order to treat such large deformations correctly, we have to evaluate the free energy of the deformed state directly without perturbations.

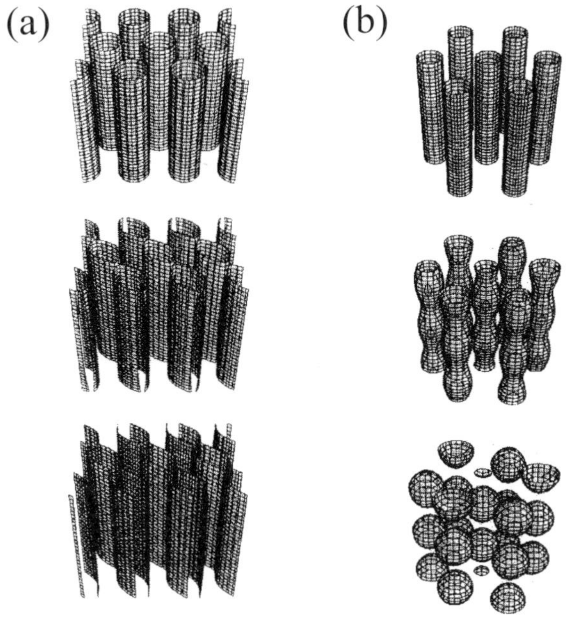

Fig. 3.32. Unstable modes of the initial cylindrical structure of an *AB* block copolymer melt (**a**) to the lamellar structure and (**b**) to the spherical structure. Reprinted with permission from M. Laradji et al., Macromolecules **30**, 3242 (1997). (© 1997, American Chemical Society)

Designing Compatibilizers

A block copolymer molecule is composed of two or more different types of segment. As different segments dissolve in different solvents, a block copolymer molecule has affinity with two or more different solvents. This is called the amphiphilic property. Amphiphilic molecules are adsorbed onto the interface when they are dissolved in a phase-separating mixture, thereby lowering the interfacial tension. This property is called the surface active effect or the compatibilizing effect. Materials that have the surface active effect are called surfactants or compatibilizers. A well known example of a compatibilizer for an oil/water mixture is the detergent. On the other hand, block copolymers play the role of compatibilizer in polymer mixtures.

As described in Sect. 3.1, polymers with high molecular weights are usually immiscible. On the other hand, most industrial polymer materials are made by blends of several polymers and fillers,[4] which should be uniform at least on mesoscopic scales, to achieve macroscopic uniformity. Therefore,

[4] Fillers are solid particles, such as carbon clusters or glass fibers, immersed in a polymer material. Such filler particles improve the mechanical, optical and electrical properties of the polymer material.

it is essential to design good compatibilizers for polymer blends. The self-consistent field theory can give important insights into the molecular design of a block copolymer as a compatibilizer, using the following degrees of freedom:

- chain length,
- type of subchain,
- architecture of the chain, such as block ratio, sequence of segments, and branching structure.

We can change these parameters in the self-consistent field theory to look for a block copolymer that has the desirable compatibilizing properties.

As an example of such studies, we discuss here the design of a compatibilizer using comb-type block copolymers. Figure 3.33 is a schematic picture of the interfacial structure of an A/B homopolymer blend containing comb-type block copolymers as compatibilizers. Here we have added two types of comb copolymer, one composed of an A-type backbone and C-type side chains, and the other composed of a B-type backbone and C-type side chains. If the chemical property of the C-type segment is in-between the A and B segments, a double layer of the AC block copolymer and the BC block copolymer is formed at the A/B homopolymer interface.

Figure 3.34 shows the predictions of the self-consistent field theory for the interfacial structure and surface tension of this mixture. Figure 3.34b shows that the longer the side chains become, the smaller is the interfacial tension. An interesting result can be observed for the case where the side chains have length 233. In this case, the interfacial tension vanishes when an appropriate amount of the block copolymer is added. This vanishing interfacial tension implies that there is no energy penalty in producing extra interfaces. The

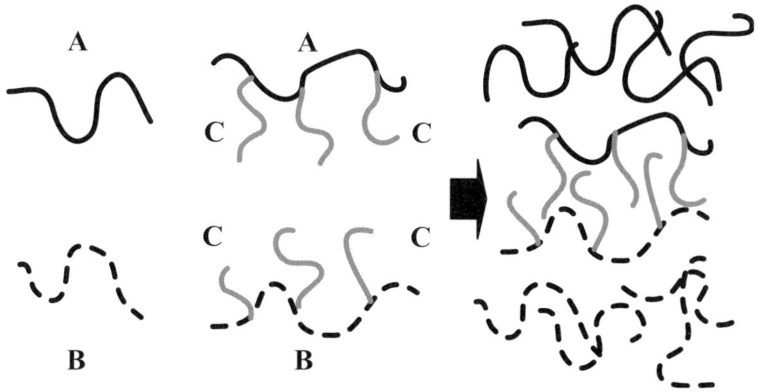

Fig. 3.33. Structure of an A/B homopolymer blend containing AC-type and BC-type comb copolymers as compatibilizers

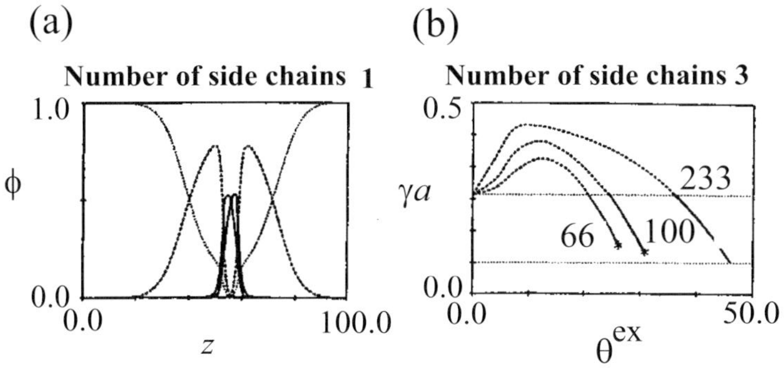

Fig. 3.34. Results of a self-consistent field calculation on an A/B homopolymer blend containing AC and BC comb copolymers as compatibilizers. (**a**) Interfacial structure in the case of a comb copolymer with only one side chain. The *curves* show, from *left to right*, the density profiles of A homopolymer, A segments and C segments of the AC block copolymer, C segments and B segments of the BC block copolymer, and B homopolymer. (**b**) Interfacial tension as a function of the concentration of the added block copolymer (θ^{ex}), shown for block copolymers with various side chain lengths. (The number of side chains is fixed at 3.) Reprinted with permission from R. Israels et al., Macromolecules **28**, 218 (1995). (© 1995, American Chemical Society)

system thus spontaneously creates more and more interfaces. In this case, the equilibrium state of the mixture possesses a fine structure with thermodynamically stable domains on mesoscopic scales. Such a phase is called the microemulsion phase. The microemulsion phase is often observed in water/oil/surfactant mixtures. Recently, a similar microemulsion phase has been found in a polymer blend containing a block copolymer as compatibilizer [37].

Exercises

Exercise 3.1. Derive the basic equations of the self-consistent field theory for an AB diblock copolymer melt. Then, choosing appropriate units of distance, chain length and energy, rewrite the equations in dimensionless form and confirm that the behavior of a block copolymer melt is determined by two parameters: the block ratio $f = N_A/(N_A + N_B)$ and χN. Hint: Denote the units of distance, chain length and energy by l_0, N_0 and ϵ_0, respectively. Then, the dimensionless forms of the equations can be obtained by introducing the dimensionless variables $\hat{\boldsymbol{r}}$, $\hat{s}$ and $\hat{F}$ by

$$\boldsymbol{r} = l_0 \hat{\boldsymbol{r}} \;, \qquad s = N_0 \hat{s} \;, \qquad F = \epsilon_0 \hat{F} \;. \tag{3.75}$$

Exercise 3.2. Consider how to calculate the statistical weight of a branched polymer such as a star polymer or a comb polymer using the path integral.

4 Ginzburg–Landau Theory

In the self-consistent field theory discussed in Chap. 3, the conformation entropy of the chains is evaluated accurately using the path integral formalism. A further simplification of the model can be made by approximating the conformational entropy and the free energy. A typical example is the Ginzburg–Landau model. In this model, the free energy is expanded in a power series in the segment density distribution, where the expansion coefficients are evaluated using the so-called random phase approximation. In this chapter, we discuss the essence of the random phase approximation and the Ginzburg–Landau model.

4.1 Formulation of the Ginzburg–Landau Theory

4.1.1 Ginzburg–Landau Model

Expansion of Free Energy and Ginzburg–Landau Model

In the self-consistent field theory discussed in Chap. 3, the conformation entropy of the polymer chains was evaluated using the path integral formalism. Such a method is suitable for the strong segregation regime where the system is divided into domains separated by sharp interfaces (see Fig 4.1). On the other hand, there is another convenient description that is valid in the vicinity of the critical point. When we approach the critical point, the difference in the average densities (compositions) between coexisting phases becomes smaller and the characteristic length scale of the density fluctuation (such as the correlation length of the density fluctuation) tends to diverge. (The density difference vanishes at the critical point, and the system is in a single-phase region above the critical temperature.) Thus, the density (composition) fluctuates only slightly around its average value. Such a region is called the weak segregation region. The small and long wavelength density modulation allows us to approximate the free energy functional by the leading terms in its functional Taylor series expansion around the average density. The model free energy obtained in this manner is called the Ginzburg–Landau model.

We will explain this model using a polymer solution as one of the simplest examples. As in the derivation of (3.51), let us assume that the free energy F

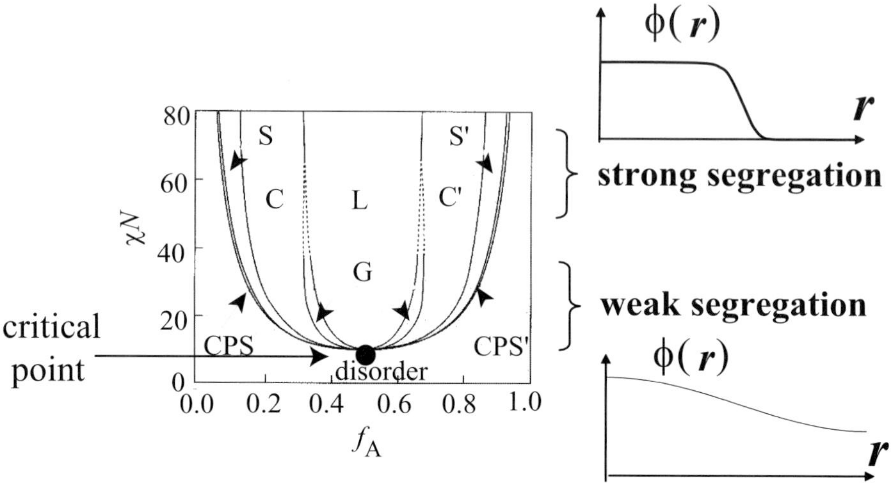

Fig. 4.1. Typical phase diagram of a block copolymer melt. The characteristic behavior of the density profile and useful theories are shown for the weak segregation region (close to the critical point denoted by the *solid circle*) and the strong segregation region (well inside the coexistence curve). Reprinted with permission from M.W. Matsen and F.S. Bates, Macromolecules **29**, 1091 (1996). (© 1996, American Chemical Society)

is a functional of the segment density distribution $\phi(\boldsymbol{r})$. Expanding the free energy $F[\phi(\boldsymbol{r})]$ with respect to the density fluctuation around the average value $\phi^{(0)}$, i.e., $\delta\phi(\boldsymbol{r}) \equiv \phi(\boldsymbol{r}) - \phi^{(0)}$, we obtain

$$F[\{\phi_K(\boldsymbol{r})\}] = F_0 + \int \mathrm{d}\boldsymbol{r} \left[a_1 \delta\phi(\boldsymbol{r}) + a_2 \left[\delta\phi(\boldsymbol{r})\right]^2 + \cdots + b_1 |\nabla \delta\phi(\boldsymbol{r})|^2 + \cdots \right] , \quad (4.1)$$

where F_0 is the free energy of a uniform system with constant density $\phi^{(0)}$, and $a_1, a_2, \ldots$ and $b_1, \ldots$ are expansion coefficients. Once such an analytic expression is obtained for the free energy, we can find the most stable structure by comparing the values of the free energy for different structures (see Chap. 3). It is well known that, not only the polymer mixtures, but also mixtures of molecules with low molecular weights show a similar phase transition and phase separation phenomena. On the level of the Ginzburg–Landau model, the difference between these two types of system arises from the difference in the number of terms retained in the expansion and the difference in the values of the expansion coefficients. As the expansion in (4.1) can be truncated for systems in the weak segregation regime, there is no essential difference in the phase transition behavior between the polymeric systems and systems composed of low-molecular-weight components. This leads to universal behavior that is independent of the microscopic detail of the constituent molecules. Such universal behavior observed near a critical point is referred to as a critical phenomenon.

Ginzburg–Landau Model and Phase Transition

In the self-consistent field theory described in Chap. 3, the value of the free energy is obtained by a numerical iteration procedure. As such iteration is not required in the Ginzburg–Landau model, we can easily obtain the phase diagram or simulate the dynamics of a system that undergoes phase transition.

The Ginzburg–Landau free energy is also useful for understanding the qualitative behavior of the phase transition. Figure 4.2 shows the typical behavior of the Ginzburg–Landau free energy for a uniform system. [When we assume homogeneity of the system, all terms containing spatial derivatives in (4.1) drop out. Then the free energy is expressed as a function (not a functional) of the average density.] Each of the minimal points of the free energy function corresponds to a stable or a metastable state. The minimum point of the free energy corresponds to the thermodynamically stable state, while the other minimal points correspond to metastable states. Although the metastable state is stable against a small perturbation, a large perturbation can induce a transition from the metastable state to another (meta)stable state that has smaller free energy. Thus, a metastable state has a finite lifetime.

Figure 4.2a shows an example of typical behavior for a continuous phase transition. If the temperature is higher than a certain temperature T_c, the free energy has only a single stable point (global minimum). The equilibrium state in the high temperature region is thus a single uniform phase. As the temperature is decreased, this stable phase starts to split into two stable phases at T_c. These two stable phases coexist below T_c and the density difference between these two phases increases further as the temperature drops. (Strictly speaking, if there is a conservation law for the constituent components of the system, the densities of the coexisting phases correspond to the two contact points of a common tangent line to the free energy curve.) In this case, there is no discontinuity in the first derivative of the free energy with respect to the density or the entropy. This is why the transition is called a continuous phase transition. The temperature T_c corresponds to the point where the difference between the two coexisting phases vanishes. At this temperature, one can show that the second and higher order derivatives of the free energy show discontinuity or divergence. Such a temperature is called the critical temperature (see the box entitled *Classification of Phase Transitions*). If the second order derivative has a discontinuity, this type of phase transition is usually called a second order phase transition.

In the case shown in Fig. 4.2b, on the other hand, the free energy function has two local minima, each of which corresponds to either a stable or a metastable phase (the A phase and the B phase). The equilibrium state is the one that has the lower free energy. When the temperature is decreased, the equilibrium state changes from the A phase to the B phase at $T = T_0$. Therefore, a discontinuous phase transition from the A phase to the B

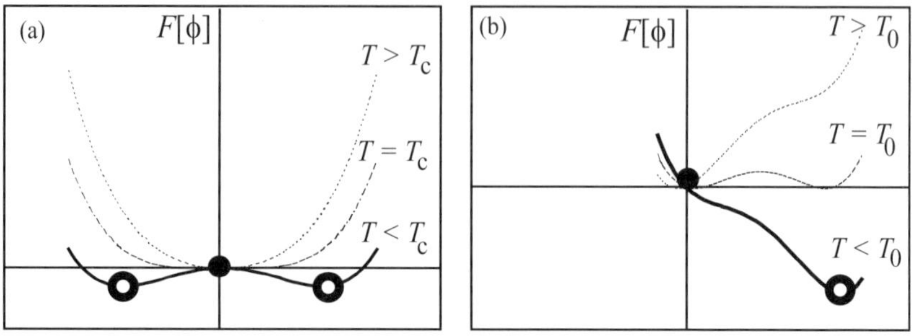

Fig. 4.2. Schematic illustrations of the Ginzburg–Landau free energy function for a uniform system as a function of the density. The characteristic behavior of the free energy function above and below the critical temperature (or transition temperature) is qualitatively different in each case. (**a**) A continuous (second order) phase transition and (**b**) a first order phase transition. The stable phase in the high temperature region is shown by a *solid circle*, while those in the low temperature region are shown by *open circles*. The temperature T_c is called the critical temperature, and T_0 is the transition temperature

phase is observed at $T = T_0$. (For example, the segment density changes discontinuously.) In this case, the phase transition is called the first order phase transition because the first derivative of the free energy with respect to the density changes discontinuously.

Classification of Phase Transitions

In thermodynamics, the canonical ensemble is characterized by the Helmholtz free energy F, whose natural variables are the temperature T, the system volume $\mathcal{V}$ and the total number of particles N. Then the total differential of F is given by

$$\mathrm{d}F = -S\mathrm{d}T - P\mathrm{d}\mathcal{V} + \mu\mathrm{d}N \,.$$

When two phases coexist, the thermodynamic forces of these two phases, such as the temperature T, pressure P and chemical potential μ, should be the same. (This property is derived from the principle of maximum entropy in the equilibrium state.) From the above expression, we can derive the relations

$$S = -\left(\frac{\partial F}{\partial T}\right)_{\mathcal{V},N} , \qquad C_{\mathcal{V}} = T\left(\frac{\partial S}{\partial T}\right)_{\mathcal{V},N} = -\left(\frac{\partial^2 F}{\partial T^2}\right)_{\mathcal{V},N} .$$

Thus, the first and second derivatives of the free energy F are the entropy S and the specific heat at constant volume $C_{\mathcal{V}}$, respectively. In a first order phase transition, the entropy shows a discontinuity ΔS, which is observed as a latent heat $T\Delta S$. In the case of a second order phase transition, the entropy is continuous at the transition point but the specific heat $C_{\mathcal{V}}$ shows a discontinuous jump.

General Expression for the Free Energy

Let us derive the expression for the free energy functional based on the Ginzburg–Landau model. Equation (3.51) derived in Sect. 3.3.2 is a general expression that can be applied to any Hamiltonian with the form

$$\hat{\mathcal{H}}(\Gamma) = \hat{\mathcal{H}}_0(\Gamma) + \hat{\mathcal{W}}[\hat{\phi}(\boldsymbol{r};\Gamma)] \,. \tag{4.2}$$

In deriving this expression, we used a system composed of independent ideal chains as the reference state $\hat{\mathcal{H}}_0$.

On the other hand, a suitable choice of the reference state $\hat{\mathcal{H}}_0$ for the Ginzburg–Landau model is a uniformly mixed state with the average density (see Fig. 4.3). [In the path integral formalism, the conformation entropy of a chain in a uniformly mixed state is the same as that in a mixture of independent ideal chains because both of them obey the Gaussian statistics of ideal chains (see Sect. 3.2.2).] For such a choice of $\hat{\mathcal{H}}_0$, we can replaced the quantities in (4.2) in the following way:

$$\begin{aligned} \phi_K(\boldsymbol{r}) &\to \delta\phi_K(\boldsymbol{r}) \equiv \phi_K(\boldsymbol{r}) - \bar{\phi}_K \qquad (\bar{\phi}_K = \text{average density}) \,, \\ \hat{\mathcal{H}}_0 &\to \text{uniformly mixed state} \,, \\ \hat{\mathcal{W}} &\to 0 \,. \end{aligned} \tag{4.3}$$

We assume here that the density fluctuation $\delta\phi_K(\boldsymbol{r})$ is induced by an (artificial) external potential $V_K(\boldsymbol{r})$ rather than being generated by the interaction between segments. Then the expression (3.51) reduces to

$$F\big[\{\delta\phi_K(\boldsymbol{r})\}\big] = -k_{\mathrm{B}}T \ln Z[\{V_K(\boldsymbol{r})\}] - \sum_K \int \mathrm{d}\boldsymbol{r} V_K(\boldsymbol{r})\delta\phi_K(\boldsymbol{r}) \,. \tag{4.4}$$

The first term on the right-hand side is the free energy as a functional of the external field $\{V_K(\boldsymbol{r})\}$. The second term is a Legendre transformation that

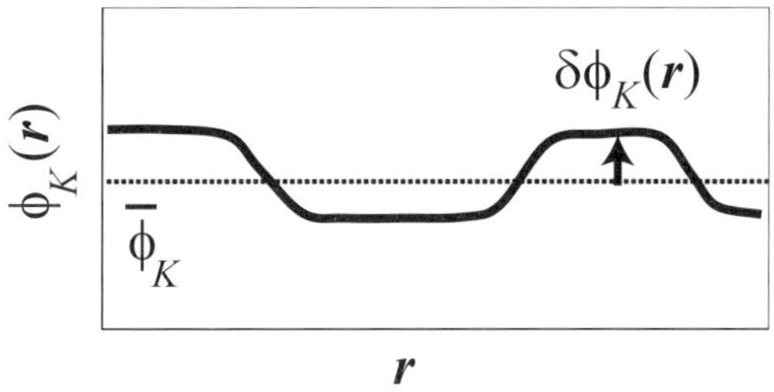

Fig. 4.3. In the Ginzburg–Landau model, we take a uniform state with density $\bar{\phi}_K$ as the reference state and consider the deviation $\delta\phi_K(\boldsymbol{r}) = \phi_K(\boldsymbol{r}) - \bar{\phi}_K$ of the density from this reference state

transforms the variable of the free energy from the external field $\{V_K(\boldsymbol{r})\}$ to its conjugate field, i.e., the density field $\{\delta\phi_K(\boldsymbol{r})\}$. In this case, the external field $\{V_K(\boldsymbol{r})\}$ is an artificial constraint force that generates the desired density profile $\{\delta\phi_K(\boldsymbol{r})\}$. The contribution from such an artificial force is cancelled by the second term of (4.4).

Density Functional Theory

In both the self-consistent field theory given in Chap. 3 and the Ginzburg–Landau theory described in this chapter, the state of the system and its free energy are described in terms of the segment density distributions. Such a theoretical framework is called a density functional theory. The density functional theory was originally developed to calculate the energy states of electrons in solids and molecules [38]. There are many similarities between the density functional theories for electrons and those for polymers.

The basic assumption underlying the density functional theories is that the free energy F is a unique functional of the density field $\phi(\boldsymbol{r})$, i.e., $F = F[\phi(\boldsymbol{r})]$, which takes its minimum value at the equilibrium state. For this reason, the basic procedure for determining the equilibrium structures and the phase diagrams is common to both the self-consistent field theory and the Ginzburg–Landau theory.

Subtracting the free energy of the reference uniform state from both sides of (4.4), we obtain the following expression for the excess free energy:

$$F\big[\{\delta\phi_K(\boldsymbol{r})\}\big] - F\big[\{0\}\big] = -k_{\mathrm{B}}T \ln \frac{Z[\{V_K(\boldsymbol{r})\}]}{Z[\{0\}]} - \sum_K \int \mathrm{d}\boldsymbol{r} V_K(\boldsymbol{r}) \delta\phi_K(\boldsymbol{r}) \,, \tag{4.5}$$

where we have used the fact that the external field $\{V_K(\boldsymbol{r})\}$ vanishes in the reference uniform state.

In the density functional theories for electronic systems, a well-known theorem called the Hohenberg–Kohn theorem tells us that there is a one-to-one mapping between the electron density distribution and the external potential field [38]. This theorem is also applicable to the present polymer case, and there is therefore a one-to-one mapping between the external constraint force $\{V_K(\boldsymbol{r})\}$ and the density field $\{\delta\phi_K(\boldsymbol{r})\}$. (We will show how this mapping is obtained later.) As a result, a small density fluctuation in the weak segregation regime means a small external field. Then we can expand the right-hand side of (4.5) in terms of the external field $\{V_K(\boldsymbol{r})\}$ instead of expanding it with respect to the segment density $\{\delta\phi_K(\boldsymbol{r})\}$. After the expansion, we rewrite $\{V_K(\boldsymbol{r})\}$ in terms of $\{\delta\phi_K(\boldsymbol{r})\}$ to obtain an expression for the free energy as a functional of $\{\delta\phi_K(\boldsymbol{r})\}$. This procedure is described in detail below.

4.1.2 Expansion of the Free Energy

Using the formula (3.34), let us expand the free energy (4.5) in a functional Taylor series around the uniformly mixed state $\{V_K(\boldsymbol{r})\} = \{0\}$. Then the first term on the right-hand side is expanded as

$$-k_{\mathrm{B}}T \ln \frac{Z[\{V_K(\boldsymbol{r})\}]}{Z[\{0\}]} = C_0 + \sum_K \int \mathrm{d}\boldsymbol{r}\, C_1^K(\boldsymbol{r}) V_K(\boldsymbol{r}) \tag{4.6}$$
$$+\frac{1}{2}\sum_K \sum_{K'} \int \mathrm{d}\boldsymbol{r} \int \mathrm{d}\boldsymbol{r}'\, C_2^{KK'}(\boldsymbol{r},\boldsymbol{r}') V_K(\boldsymbol{r}) V_{K'}(\boldsymbol{r}') + \cdots ,$$

where the expansion coefficients $C_n^{K,K',\ldots}(\boldsymbol{r},\boldsymbol{r}',\ldots)$ are defined by

$$C_n^{K,K',\ldots}(\boldsymbol{r},\boldsymbol{r}',\ldots) \equiv -k_{\mathrm{B}}T \frac{\delta^n}{\delta V_K(\boldsymbol{r}) \delta V_{K'}(\boldsymbol{r}') \ldots} \ln \left. \frac{Z[\{V_K\}]}{Z[\{0\}]} \right|_{\{V_K\}=\{0\}} . \tag{4.7}$$

Equation (4.6) is a power series expansion of the logarithm of the partition function, i.e., the free energy. The coefficients of such an expansion are called cumulants. The box entitled *Definitions and Physical Meaning of Cumulants* gives explicit expressions for the cumulants and their physical meaning.

Substituting (4.14) into (4.6), we obtain the following expression for $\ln Z[\{V_K(\boldsymbol{r})\}]$:

$$-k_{\mathrm{B}}T \ln \frac{Z[\{V_K(\boldsymbol{r})\}]}{Z[\{0\}]} = -\frac{\beta}{2} \sum_K \sum_{K'} \int \mathrm{d}\boldsymbol{r} \int \mathrm{d}\boldsymbol{r}'\, S_{KK'}^{(\mathrm{u})}(\boldsymbol{r}-\boldsymbol{r}') V_K(\boldsymbol{r}) V_{K'}(\boldsymbol{r}') + \cdots . \tag{4.8}$$

Using this relation, we can obtain an explicit expression for the segment density $\{\delta\phi_K(\boldsymbol{r})\}$ in terms of the external potential field $\{V_K(\boldsymbol{r})\}$. For this purpose, we use a similar relation to the second equation of (3.50):

$$\delta\phi_K(\boldsymbol{r}) = -k_{\mathrm{B}}T \frac{\delta}{\delta V_K(\boldsymbol{r})} \ln \frac{Z[\{V_K\}]}{Z[\{0\}]}$$
$$= -\beta \sum_{K'} \int \mathrm{d}\boldsymbol{r}'\, S_{KK'}^{(\mathrm{u})}(\boldsymbol{r}-\boldsymbol{r}') V_{K'}(\boldsymbol{r}') + \cdots . \tag{4.9}$$

This relation can be inverted to give

$$-\beta V_K(\boldsymbol{r}) = \sum_{K'} \int \mathrm{d}\boldsymbol{r}' \left(S^{(\mathrm{u})}\right)^{-1}_{KK'}(\boldsymbol{r}-\boldsymbol{r}') \delta\phi_{K'}(\boldsymbol{r}') , \tag{4.10}$$

where $\left(S^{(\mathrm{u})}\right)^{-1}_{KK'}$ is the inverse of the two-point correlation function $S_{KK'}^{(\mathrm{u})}$ defined by

$$\sum_{K''} \int \mathrm{d}\boldsymbol{r}''\, S_{KK''}^{(\mathrm{u})}(\boldsymbol{r}-\boldsymbol{r}'') \left(S^{(\mathrm{u})}\right)^{-1}_{K''K'}(\boldsymbol{r}''-\boldsymbol{r}') = \delta_{KK'} \delta(\boldsymbol{r}-\boldsymbol{r}') . \tag{4.11}$$

Definition and Physical Meaning of Cumulants

The partition function $Z[\{V_K\}]/Z[\{0\}]$ on the right-hand side of (4.6) is rewritten using the definitions (3.38), (3.47) and $\hat{\mathcal{W}} = 0$ as

$$\frac{Z[\{V_K(\boldsymbol{r})\}]}{Z[\{0\}]} = \left\langle \exp\Big[-\beta \sum_K \int V_K(\boldsymbol{r})\hat{\phi}_K(\boldsymbol{r};\Gamma)\mathrm{d}\boldsymbol{r}\Big]\right\rangle_0 , \tag{4.12}$$

where $\langle \ * \ \rangle_0$ is the ensemble average over the equilibrium canonical distribution of the reference uniform state, defined by

$$\begin{aligned}\left\langle \ * \ \right\rangle_0 &\equiv \frac{1}{Z[\{0\}]}\int \mathrm{d}\Gamma \ * \ \exp\big[-\beta\hat{\mathcal{H}}_0(\Gamma)\big] \\ &= \frac{\int \mathrm{d}\Gamma \ * \ \exp\big[-\beta\hat{\mathcal{H}}_0(\Gamma)\big]}{\int \mathrm{d}\Gamma \exp\big[-\beta\hat{\mathcal{H}}_0(\Gamma)\big]} .\end{aligned} \tag{4.13}$$

Using this definition, (4.7) is rewritten as

$$\begin{aligned} C_0 &= 0 , \\ C_1^K(\boldsymbol{r}) &= \left\langle \delta\hat{\phi}_K(\boldsymbol{r};\Gamma)\right\rangle_0 = 0 , \\ C_2^{KK'}(\boldsymbol{r},\boldsymbol{r}') &= -\beta\left\langle \delta\hat{\phi}_K(\boldsymbol{r};\Gamma)\delta\hat{\phi}_{K'}(\boldsymbol{r}';\Gamma)\right\rangle_0 \equiv -\beta S^{(\mathrm{u})}_{KK'}(\boldsymbol{r}-\boldsymbol{r}') ,\end{aligned} \tag{4.14}$$

and so on, where $S^{(\mathrm{u})}_{KK'}(\boldsymbol{r})$ is the two-point correlation function of the segment density fluctuations of the reference uniform state. Note that these quantities are functions of $\boldsymbol{r} - \boldsymbol{r}'$ due to the translational invariance of the uniform system.

The expansion in (4.6) is a power series expansion of the logarithm of the partition function. Such an expansion is called a cumulant expansion, and the expansion coefficients are called cumulants. The quantities C_n in (4.14) are examples of the lower order cumulants. As shown in (4.14), first and second order cumulants coincide with first and second order correlation functions of the fluctuation. Higher order cumulants, however, have more complex forms that also depend on lower order correlation functions. In general, the n th order cumulant expresses the net n-body correlation that cannot be described by any products of the correlation functions up to $(n-1)$ th order.

Substituting (4.8) and (4.10) into (4.5), we finally obtain the expression:

$$
\begin{aligned}
&F[\{\delta\phi_K(\boldsymbol{r})\}] - F[\{0\}] \\
&= \frac{1}{(2\pi)^d}\frac{1}{2\beta}\sum_{KK'}\int \mathrm{d}\boldsymbol{r}\int \mathrm{d}\boldsymbol{r}'\left(S^{(\mathrm{u})}\right)^{-1}_{KK'}(\boldsymbol{r}-\boldsymbol{r}')\delta\phi_K(\boldsymbol{r})\delta\phi_{K'}(\boldsymbol{r}') + \cdots \\
&= \frac{1}{2\beta}\sum_{KK'}\sum_{\boldsymbol{q}}\left(\tilde{S}^{(\mathrm{u})}\right)^{-1}_{KK'}(\boldsymbol{q})\delta\tilde{\phi}_K(\boldsymbol{q})\delta\tilde{\phi}_{K'}(-\boldsymbol{q}) + \cdots , \qquad (4.15)
\end{aligned}
$$

where d is the dimensionality of the system. The first line of (4.15) is an expression in the real space and the second line is its Fourier space representation. Note that the expansion coefficients in the Fourier space representation are related to the scattering functions which are the Fourier transforms of the real-space correlation functions (see Sect. 2.3.2).

Expansion Coefficients and Scattering Functions

In the Ginzburg–Landau expansion in (4.15), the coefficient of the second order term, $\tilde{S}^{(\mathrm{u})}(\boldsymbol{q})$, is the inverse of the power spectrum of the segment density fluctuation $\delta\tilde{\phi}_K(\boldsymbol{q})$, i.e., the Fourier transform of the two-point correlation function of the density fluctuations. This can be understood intuitively as follows.

For simplicity, we consider a one-component system and we truncate the expansion in (4.15) at the second order term. Using this free energy, we can obtain the canonical ensemble of the density fluctuation $\delta\tilde{\phi}(\boldsymbol{q})$ as

$$
P(\delta\tilde{\phi}(\boldsymbol{q})) = \text{const.} \times \exp\{-\beta F[\delta\tilde{\phi}(\boldsymbol{q})]\} = \text{const.} \times \exp\left[-\frac{1}{2}\sum_{\boldsymbol{q}}\frac{|\delta\tilde{\phi}(\boldsymbol{q})|^2}{\tilde{S}^{(\mathrm{u})}(\boldsymbol{q})}\right] ,
$$

where the constant is the normalization constant. The variance of this probability distribution for $\delta\tilde{\phi}(\boldsymbol{q})$ is given by

$$
\langle|\delta\tilde{\phi}(\boldsymbol{q})|^2\rangle = \tilde{S}^{(\mathrm{u})}(\boldsymbol{q}) .
$$

The left-hand side is the Fourier transform of the correlation function of the density fluctuation, i.e., the scattering function introduced in Sect. 2.3.2. Hence, the above relation shows that the coefficient of the second order term of the free energy expansion in Fourier space is the inverse of the scattering function, i.e., the Fourier transform of the two-point correlation function. In a similar manner, we can also demonstrate that coefficients of the higher order terms can be expressed in terms of the many-body correlation functions. Therefore, expanding the free energy corresponds to calculating many-body correlation functions of the interacting many-body system.

Equation (4.15) is a formal expansion of the free energy in terms of the density fluctuations. Truncating this expansion at terms of the appropriate order gives the Ginzburg–Landau free energy. The nature of the phase transition can be found from the sign of the eigenvalues of the coefficient matrix of the second order term in (4.15). (Such a matrix is the second derivative of the free energy functional with respect to the segment density. A second derivative matrix of this kind is called a Hessian.) If all the eigenvalues of the matrix $\left(\tilde{S}^{(\mathrm{u})}\right)^{-1}_{KK'}(\boldsymbol{q})$ are positive for any wave numbers $\boldsymbol{q}$, any small density fluctuation around the uniform state causes an increase in the free energy. The uniform state is then thermodynamically stable against small fluctuations. On the other hand, if at least one of the eigenvalues of $\left(\tilde{S}^{(\mathrm{u})}\right)^{-1}_{KK'}(\boldsymbol{q})$ is negative for a certain wave number $|\boldsymbol{q}^*|$, the uniform state is unstable and the fluctuation with wave number $|\boldsymbol{q}^*|$ grows. The point where the instability of the uniform state first occurs is called the spinodal point.

4.1.3 Evaluation of Expansion Coefficients Using the Random Phase Approximation

Linear Response Theory

The unknown parameters in (4.15) are the inverse of the correlation functions of the segment density fluctuation in the uniform state $S^{(\mathrm{u})}_{KK'}(\boldsymbol{r})$. The architectures of the polymer chains are reflected in the Ginzburg–Landau free energy through these correlation functions. In this section, we show how to obtain explicit expressions for these correlation functions using the random phase approximation (RPA). The RPA is a technique to calculate the correlation functions of interacting polymer systems based on the linear response theory (see Fig. 4.4).

To explain the RPA method, we use a one-component polymer solution as a simple example. Let us decompose the Hamiltonian of this mutually

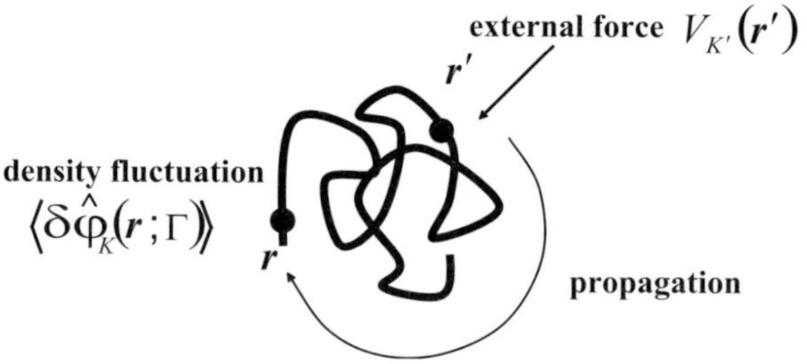

Fig. 4.4. The concept of the random phase approximation. The external force $V_{K'}(\boldsymbol{r}')$ imposed at position $\boldsymbol{r}'$ induces a density fluctuation $\langle\delta\phi_K(\boldsymbol{r})\rangle$ at a distant position $\boldsymbol{r}$

interacting chain system in the same manner as (3.37). We introduce an ideal system composed of non-interacting ideal chains and use it as the reference state $\hat{\mathcal{H}}_0(\Gamma)$. Due to the translational symmetry, this reference system is uniform unless any external field is imposed. We then impose an artificial external potential $V(\boldsymbol{r})$ to induce a deviation of the segment density $\delta\hat{\phi}(\boldsymbol{r};\Gamma)$ from the uniform reference value ϕ_0:

$$\hat{\phi}(\boldsymbol{r};\Gamma) = \phi_0 + \delta\hat{\phi}(\boldsymbol{r};\Gamma) \ . \tag{4.16}$$

When the external field $V(\boldsymbol{r})$ is small, the linear response theory gives the following linear relation between $\langle\delta\hat{\phi}(\boldsymbol{r};\Gamma)\rangle$ and $V(\boldsymbol{r})$ (see the box entitled *Linear Response Theory*):

$$\begin{aligned}\delta\phi(\boldsymbol{r}) &\equiv \langle\delta\hat{\phi}(\boldsymbol{r};\Gamma)\rangle \\ &= -\beta\int \mathrm{d}\boldsymbol{r}'\langle\delta\hat{\phi}(\boldsymbol{r};\Gamma)\delta\hat{\phi}(\boldsymbol{r}';\Gamma)\rangle_0 V(\boldsymbol{r}') \\ &= -\beta\int \mathrm{d}\boldsymbol{r}' S^{(0)}(\boldsymbol{r}-\boldsymbol{r}')V(\boldsymbol{r}') \ ,\end{aligned} \tag{4.17}$$

where $\langle\ \rangle_0$ and $\langle\ \rangle$ are the equilibrium ensemble average for the reference state without and with the external field V, respectively, and $S^{(0)}(\boldsymbol{r}) = \langle\delta\hat{\phi}(\boldsymbol{r};\Gamma)\delta\hat{\phi}(\boldsymbol{r}';\Gamma)\rangle_0$ is the two-point correlation function in the reference state. In the derivation of the first equality in (4.17), we used the property $\langle\delta\hat{\phi}(\boldsymbol{r};\Gamma)\rangle_0 = 0$ for the uniform reference state. To derive the second equality, we used the fact that $\langle\delta\hat{\phi}(\boldsymbol{r};\Gamma)\delta\hat{\phi}(\boldsymbol{r}';\Gamma)\rangle_0$ is a function of $\boldsymbol{r}-\boldsymbol{r}'$ due to the translational symmetry of the uniform state.

Equation (4.17) indicates that a weak external field $V(\boldsymbol{r}')$ imposed at position $\boldsymbol{r}'$ induces a segment density fluctuation at a distant point $\boldsymbol{r}$. The correlation function $S^{(0)}(\boldsymbol{r}-\boldsymbol{r}') \equiv \langle\delta\hat{\phi}(\boldsymbol{r};\Gamma)\delta\hat{\phi}(\boldsymbol{r}';\Gamma)\rangle_0$ describes how the effect of the external field at $\boldsymbol{r}'$ is propagated to $\boldsymbol{r}$. The linear perturbation theory that leads to (4.17) is called the linear response theory. (In a linear relation between two small quantities induced by a perturbation, its proportionality coefficient can be evaluated in the reference state without the perturbation.) The relation (4.17) is a general relation that can be applied, not only to polymer systems, but also to many other systems such as spin systems or phase-separating fluids. The polymeric nature is taken into account through the two-point correlation function $S^{(0)}(\boldsymbol{r})$. As we chose a mixture of non-interacting ideal chains as the reference system, the correlation function $S^{(0)}(\boldsymbol{r})$ in this reference state can easily be calculated using the Gaussian statistics described in Sect. 2.3.1.

Random Phase Approximation

We calculate the two-point correlation functions for multicomponent polymer mixtures based on the RPA, which relies on the linear relation (4.17). Let us consider the potential field acting on a K-type segment at position $\boldsymbol{r}$. In addition to the externally imposed field $V_K(\boldsymbol{r})$, the segment feels an extra potential induced by the local fluctuation in the segment density and local incompressibility condition. We denote the latter contribution by $V_K^{(\mathrm{RPA})}(\boldsymbol{r})$. The observed density fluctuation $\delta\phi_K(\boldsymbol{r})$ is a consequence of all these external fields. Taking these external potentials into account, (4.17) for a multicomponent system is rewritten as

$$\delta\phi_K(\boldsymbol{r}) = -\beta \sum_{K'} \int \mathrm{d}\boldsymbol{r}' S_{KK'}^{(0)}(\boldsymbol{r}-\boldsymbol{r}') \left[V_{K'}(\boldsymbol{r}') + V_{K'}^{(\mathrm{RPA})}(\boldsymbol{r}')\right] . \tag{4.18}$$

The convolution integral in this equation can be decomposed by Fourier transforming both sides:

$$\delta\tilde{\phi}_K(\boldsymbol{q}) = -\beta \sum_{K'} \tilde{S}_{KK'}^{(0)}(\boldsymbol{q}) \left[\tilde{V}_{K'}(\boldsymbol{q}) + \tilde{V}_{K'}^{(\mathrm{RPA})}(\boldsymbol{q})\right] . \tag{4.19}$$

This equation implies that Fourier modes with different wave number $\boldsymbol{q}$ are decoupled.

Linear Response Theory

Consider a reference state with a uniform segment density ϕ_0. We impose an external field $V(\boldsymbol{r})$ as a perturbation that induces a density fluctuation. The average of the density fluctuation $\delta\hat{\phi}(\boldsymbol{r};\Gamma)$ in the perturbed system, denoted by $\delta\phi(\boldsymbol{r}) \equiv \langle\delta\hat{\phi}(\boldsymbol{r};\Gamma)\rangle$, is given by

$$\begin{aligned}\delta\phi(\boldsymbol{r}) &\equiv \langle\delta\hat{\phi}(\boldsymbol{r};\Gamma)\rangle \\ &= \frac{1}{Z}\int \mathrm{d}\Gamma\, \delta\hat{\phi}(\boldsymbol{r};\Gamma) \exp\left\{-\beta\left[\hat{\mathcal{H}}_0(\Gamma) + \int \mathrm{d}\boldsymbol{r}' V(\boldsymbol{r}')\delta\hat{\phi}(\boldsymbol{r}';\Gamma)\right]\right\} ,\end{aligned} \tag{4.20}$$

where Z is the partition function of the perturbed system defined by

$$Z = \int \mathrm{d}\Gamma \exp\left\{-\beta\left[\hat{\mathcal{H}}_0(\Gamma) + \int \mathrm{d}\boldsymbol{r}' V(\boldsymbol{r}')\delta\hat{\phi}(\boldsymbol{r}';\Gamma)\right]\right\} . \tag{4.21}$$

If the perturbation term $-\beta\int \mathrm{d}\boldsymbol{r} V(\boldsymbol{r})\delta\hat{\phi}(\boldsymbol{r};\Gamma)$ is small, we can linearize (4.20) with respect to this small perturbation as

$$\begin{aligned}&\exp\left\{-\beta\left[\hat{\mathcal{H}}_0(\Gamma) + \int \mathrm{d}\boldsymbol{r}' V(\boldsymbol{r}')\delta\hat{\phi}(\boldsymbol{r}';\Gamma)\right]\right\} \\ &\qquad \sim \exp\left[-\beta\hat{\mathcal{H}}_0(\Gamma)\right]\left[1 - \beta\int \mathrm{d}\boldsymbol{r}' V(\boldsymbol{r}')\delta\hat{\phi}(\boldsymbol{r}';\Gamma)\right] ,\end{aligned} \tag{4.22}$$

which leads to the linear response relation

$$\delta\phi(\boldsymbol{r}) = \langle\delta\hat{\phi}(\boldsymbol{r};\Gamma)\rangle = -\beta\int \mathrm{d}\boldsymbol{r}' \langle\delta\hat{\phi}(\boldsymbol{r};\Gamma)\delta\hat{\phi}(\boldsymbol{r}';\Gamma)\rangle_0 V(\boldsymbol{r}') . \tag{4.23}$$

There are two main contributions to the induced external field $V_K^{(\mathrm{RPA})}(\boldsymbol{r})$. One is the direct interaction between neighboring segments and the other is the static pressure arising from the incompressibility condition. (For polyelectrolytes, i.e., ionized polymers, a long-range Coulomb interaction should also be added. We neglect such complications in this book.) When we discussed the self-consistent field theory in Sect. 3.2.1, we assumed a strict incompressibility condition in (3.1), using the Lagrangian multiplier $\gamma(\boldsymbol{r})$. In this section, we introduce a slight compressibility instead of introducing the Lagrange multiplier (see the box entitled *Incompressibility Condition and Lagrangian Multiplier*).

Incompressibility Condition and Lagrangian Multiplier

In density functional theories such as the self-consistent field theory and the Ginzburg–Landau theory, the incompressibility condition plays an important role. In a multicomponent system, the sum of the density fluctuations over all components should vanish, i.e., $\sum_K \delta\phi_K(\boldsymbol{r}) = 0$, due to the total incompressibility condition. In order to introduce such an incompressibility condition into a reference system that does not have one, there are two main methods, each of which is realized by adding an extra term to the free energy F_0 of the reference system:

- In the method using a Lagrangian multiplier, the free energy is given by

$$F = F_0 + \int \mathrm{d}\boldsymbol{r} \sum_K \delta\phi_K(\boldsymbol{r})\gamma(\boldsymbol{r}) ,$$

  where $\gamma(\boldsymbol{r})$ is the Lagrangian multiplier which corresponds to the static pressure.
- In the method using weak compressibility, the free energy is given by

$$F = F_0 + \frac{1}{2\kappa} \int \mathrm{d}\boldsymbol{r} \left[\sum_K \delta\phi_K(\boldsymbol{r}) \right]^2 ,$$

  where κ is the compressibility of the total segment density. In the limit $\kappa \to 0$, this condition effectively reduces to the above strict incompressibility condition.

The appropriate form of the elastic energy due to the compressibility for the total density is written as

$$W_{\mathrm{elastic}} = \frac{1}{2\kappa\phi_0^2} \left[\sum_K \delta\phi_K(\boldsymbol{r}) \right]^2 . \tag{4.24}$$

Then a density fluctuation $\delta\phi_K(\boldsymbol{r})$ induces an elastic force

$$\frac{\delta}{\delta\phi_K(\boldsymbol{r})} W_{\text{elastic}} = \frac{1}{\kappa\phi_0^2} \sum_{K'} \delta\phi_{K'}(\boldsymbol{r}) \, . \tag{4.25}$$

Adding this contribution to the segment interaction and Fourier transforming the result, we obtain

$$\tilde{V}_K^{(\text{RPA})}(\boldsymbol{q}) = \sum_{K'} \left(\chi_{KK'} + \frac{1}{\kappa\phi_0^2} \right) \delta\tilde{\phi}_{K'}(\boldsymbol{q}) \equiv \sum_{K'} \hat{\chi}_{KK'} \delta\tilde{\phi}_{K'}(\boldsymbol{q}) \, , \tag{4.26}$$

where the new parameter $\hat{\chi}_{KK'}$ is defined as the sum of the interaction parameter $\chi_{KK'}$ and the compressibility coefficient.

Substituting (4.26) into (4.19) and collecting the terms containing $\delta\tilde{\phi}_{K'}(\boldsymbol{q})$ on the left-hand side, we obtain

$$\sum_{K'} \tilde{A}_{KK'}^{-1}(\boldsymbol{q}) \delta\tilde{\phi}_{K'}(\boldsymbol{q}) = -\beta \sum_{K'} \tilde{S}_{KK'}^{(0)}(\boldsymbol{q}) \tilde{V}_{K'}(\boldsymbol{q}) \, , \tag{4.27}$$

where we have introduced the matrix

$$\tilde{A}_{KK'}(\boldsymbol{q}) = \delta_{KK'} + \beta \sum_{K''} \tilde{S}_{KK''}^{(0)}(\boldsymbol{q}) \hat{\chi}_{K''K'} \, . \tag{4.28}$$

Multiplying both sides by the inverse matrix of $\tilde{A}_{KK'}(\boldsymbol{q})$, we obtain

$$\delta\tilde{\phi}_K(\boldsymbol{q}) = -\beta \sum_{K'} \left[\sum_{K''} \tilde{A}_{KK''}^{-1}(\boldsymbol{q}) \tilde{S}_{K''K'}^{(0)}(\boldsymbol{q}) \right] \tilde{V}_{K'}(\boldsymbol{q}) \, , \tag{4.29}$$

where $\tilde{A}_{KK'}^{-1}$ is the KK' element of the inverse matrix of $\tilde{A}_{KK'}$. [The relation that connects $\tilde{A}_{KK'}(\boldsymbol{q})$ and $\tilde{A}_{KK'}^{-1}(\boldsymbol{q})$ is obtained by Fourier transforming (4.11) as $\sum_{K''} \tilde{A}_{KK''}(\boldsymbol{q}) \tilde{A}_{K''K'}^{-1}(\boldsymbol{q}) = \delta_{KK'}$.]

As (4.29) gives the linear relation between the externally imposed field $\tilde{V}_{K'}(\boldsymbol{q})$ and the induced response $\langle \delta\tilde{\phi}_K(\boldsymbol{q}) \rangle$, the linear response relation (4.17) tells us that the proportionality coefficient in (4.29) corresponds to the Fourier transform of the correlation function. We thus obtain

$$\tilde{S}_{KK'}^{(\text{u})}(\boldsymbol{q}) = \sum_{K''} \tilde{A}_{KK''}^{-1}(\boldsymbol{q}) \tilde{S}_{K''K'}^{(0)}(\boldsymbol{q}) \, . \tag{4.30}$$

As the scattering function of the reference ideal system $\tilde{S}_{KK'}^{(0)}(\boldsymbol{q})$ has already been given in (2.42), we have only to calculate $\tilde{A}_{KK'}^{-1}$.

Although we have only discussed the two-body correlation functions, the same technique can be applied to many-body correlations as well. One of the advantages of the RPA for polymer systems is that it can be used to treat arbitrary chain architectures and arbitrary compositions analytically. This feature is in contrast to the self-consistent field theory which can only be solved numerically in general.

Finally, we must comment on the RPA calculation for the case where there is a strict incompressibility condition. In this case, the above procedure should be modified. For the case of an A/B binary system, the incompressibility condition imposes the constraint $\delta\hat{\phi}_A(\boldsymbol{r};\Gamma) + \delta\hat{\phi}_B(\boldsymbol{r};\Gamma) = 0$. (This relation holds only on a large scale. On the length scale of the segment size, the sum of the microscopic segment density does fluctuate, giving a short-range contribution to the scattering function. We neglect such short-range contributions here.) Substituting this constraint into (4.14), we obtain the relations

$$\tilde{S}^{(\mathrm{u})}_{AA}(\boldsymbol{q}) = \tilde{S}^{(\mathrm{u})}_{BB}(\boldsymbol{q}) = -\tilde{S}^{(\mathrm{u})}_{AB}(\boldsymbol{q}) \equiv \tilde{S}^{(\mathrm{u})}(\boldsymbol{q}) \,. \tag{4.31}$$

In this case, the matrix $\tilde{A}_{KK'}(\boldsymbol{q})$ defined by (4.28) becomes singular,[1] leading to the absence of $\tilde{A}^{-1}_{KK'}(\boldsymbol{q})$. In this case, we have to introduce a Lagrange multiplier $\gamma(\boldsymbol{q})$ to guarantee the incompressibility condition instead of the elastic energy given in (4.24). Then (4.27) is solved using the incompressibility condition to give an explicit expression for $\gamma(\boldsymbol{q})$. Substituting this back into (4.27) and eliminating one of the segment density fields, we finally obtain a similar equation to (4.29), from which the scattering functions can be obtained.

Examples of the Random Phase Approximation

As an example of calculations using RPA, we show the scattering functions for an A/B homopolymer blend and for an AB block copolymer melt.

- Consider an incompressible A/B homopolymer blend. If we assume a strict incompressibility condition $\delta\hat{\phi}_A(\boldsymbol{r};\Gamma) + \delta\hat{\phi}_B(\boldsymbol{r};\Gamma) = 0$, the scattering function defined in (4.31) is obtained using the RPA as

$$\frac{1}{\tilde{S}^{(\mathrm{u})}(q)} = \frac{1}{\tilde{S}^{(0)}_{AA}(q)} + \frac{1}{\tilde{S}^{(0)}_{BB}(q)} - 2\chi \,, \tag{4.32}$$

  where $\tilde{S}^{(0)}_{KK}(q)$ is the Debye function given by (2.42), which describes the scattering function of a single ideal chain. In deriving this equation, we used the fact that all the scattering functions depend on the wave number $\boldsymbol{q}$ only through its magnitude $q \equiv |\boldsymbol{q}|$ because of the isotropic symmetry of the Hamiltonian. We also used the fact that $\tilde{S}^{(0)}_{AB}(q) = 0$ holds because there are no chemical bonds between the A segments and B segments and therefore no correlations appear in the reference ideal state.
  In Fig. 4.5, the behavior of the scattering function given by (4.32) is shown for a symmetric A/B homopolymer blend with the composition

[1] One of the eigenvalues of the matrix $\tilde{A}_{KK'}(\boldsymbol{q})$ vanishes because a sum of the elements in any column equals zero due to the incompressibility condition.

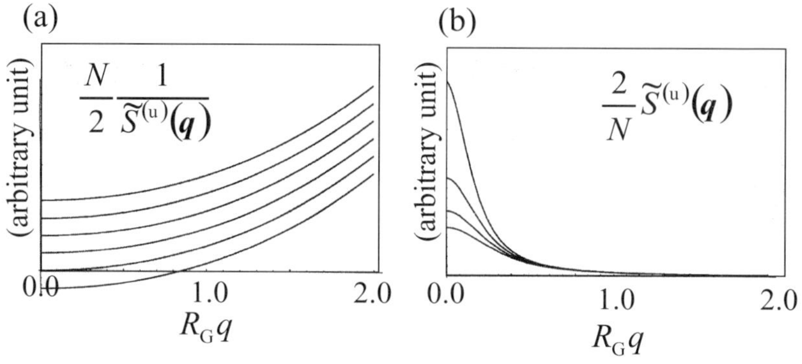

Fig. 4.5. Results of the random phase approximation on a symmetric A/B polymer blend. (**a**) Inverse of the scattering function $1/\tilde{S}^{(u)}(\boldsymbol{q})$ and (**b**) the functional form of the scattering function $\tilde{S}^{(u)}(\boldsymbol{q})$. The value of χN is chosen *from top to bottom* as $\chi N = 0.0, 0.5, 1.0, 1.5, 2.0$ and 2.5 in (**a**), and $\chi N = 1.5, 1.0, 0.5$ and 0.0 in (**b**). The critical point is located at $\chi N = 2.0$

$\bar{\phi}_A : \bar{\phi}_B = 1 : 1$. When we increase χN starting from the uniformly mixed state (i.e., decreasing the temperature from the high temperature region), $\tilde{S}^{(u)}(q)$ for $q = 0$ changes its sign at $\chi N = 2$, at which point (4.32) diverges. The negative value of $\tilde{S}^{(u)}(q = 0)$ means that any small density fluctuations with $q = 0$ decrease the free energy, leading to a growth of long wavelength fluctuations in the density field. Therefore, the uniform state is no longer stable for $\chi N > 2$. This corresponds to a macrophase separation between the A-rich phase and the B-rich phase (i.e., spinodal decomposition).

Substituting (4.32) into the first term of (4.15), we obtain the Ginzburg–Landau free energy model up to second order in the segment density fluctuations. Using the inverse Fourier transform, we obtain a real-space representation of the free energy that includes convolution integrals connecting density fluctuations between different points. A simpler and more tractable model can be obtained by expanding (4.32) in a power series in q and retaining only the terms up to the second order in q. This procedure leads to a free energy expression containing a square of the spatial derivative of the density fluctuations $\delta\phi_K(\boldsymbol{r})$, i.e., $D_{KK'}\nabla\delta\rho_K(\boldsymbol{r})\cdot\nabla\rho_{K'}(\boldsymbol{r})$. Such a term is called the square gradient energy. Combining this square gradient energy with the Flory–Huggins free energy for a uniform system given by (2.67), we obtain the so-called Flory–Huggins–de Gennes model. The Flory–Huggins–de Gennes model has been successfully applied to many polymer phase separation phenomena (see the box entitled *Flory–Huggins–de Gennes model*).

Flory–Huggins–de Gennes Model

Using the scattering functions obtained with the RPA, we can derive a Ginzburg–Landau-type free energy functional for inhomogeneous systems. As an example, we derive here a Ginzburg–Landau-type free energy for an A/B polymer blend.

First we expand $\left(S^{(\mathrm{u})}\right)^{-1}_{KK'}(q)$ in (4.15) in a Taylor series in q and retain only the terms up to second order in q. The inverse Fourier transform of this result gives

$$F = \int \mathrm{d}\boldsymbol{r} \left[f(\{\delta\phi_K(\boldsymbol{r})\}) + \frac{1}{2}\sum_{K}\sum_{K'} D_{KK'} \nabla\delta\phi_K(\boldsymbol{r}) \cdot \nabla\delta\phi_{K'}(\boldsymbol{r}) \right] , \quad (4.33)$$

where $D_{KK'}$ is the coefficient of the q^2 term in the Taylor series expansion of (4.32), and $f(\{\delta\phi_K(\boldsymbol{r})\})$ is the local potential, depending only on the local density $\delta\phi_K(\boldsymbol{r})$. As f describes the free energy of a uniform system, it is reasonable to use the Flory–Huggins free energy given by (2.67) for this local potential f. We then obtain the free energy

$$F = \int \mathrm{d}\boldsymbol{r} \left\{ \frac{\phi_A(\boldsymbol{r})}{N_A} \ln \phi_A(\boldsymbol{r}) + \frac{\phi_B(\boldsymbol{r})}{N_B} \ln \phi_B(\boldsymbol{r}) + \chi \phi_A(\boldsymbol{r}) \phi_B(\boldsymbol{r}) + \frac{b^2}{36\bar{\phi}_A\bar{\phi}_B} \left[|\nabla\delta\phi_A(\boldsymbol{r})|^2 + |\nabla\delta\phi_B(\boldsymbol{r})|^2 \right] \right\} . \quad (4.34)$$

This is an extension of the Flory–Huggins model taking the spatial variation of the polymer density into account. This free energy model is known as the Flory–Huggins–de Gennes model. This model can be used for calculations concerning inhomogeneous systems containing interfaces and domains, such as the phase separation phenomena of polymer systems.

- Considering now an incompressible AB diblock copolymer melt, the relations (4.31) are also satisfied in this case. We thus obtain the result

$$\frac{1}{\tilde{S}^{(\mathrm{u})}(q)} = \frac{4h(1,x)/N}{4h(f,x)h(1-f,x) - \left[h(1,x) - h(f,x) - h(1-f,x)\right]^2} - 2\chi , \quad (4.35)$$

where N is the total length of the block copolymer chain, x is defined by $x = (1/6)Nb^2|\boldsymbol{q}|^2$, $f = N_A/(N_A + N_B)$ is the block ratio, and $h(f,x)$ is a modified version of the Debye function defined by

$$h(f,x) = \frac{2}{x^2}\left(\mathrm{e}^{-fx} - 1 + fx\right) . \quad (4.36)$$

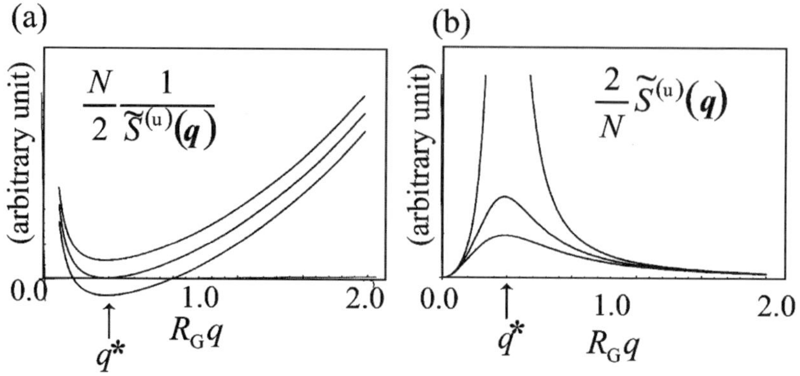

Fig. 4.6. Results of the random phase approximation on a symmetric AB diblock copolymer melt. (**a**) Inverse of the scattering function $1/\tilde{S}^{(\mathrm{u})}(\boldsymbol{q})$ and (**b**) functional form of the scattering function $\tilde{S}^{(\mathrm{u})}(\boldsymbol{q})$. The value of χN is chosen *from top to bottom* as $\chi N = 0.0, 10.0$ and 20.0 in (**a**), and $\chi N = 10.0, 5.0$ and 0.0 in (**b**). The critical point for the microphase separation is located around $\chi N = 10.5$, where the mode with $q = q^*$ becomes unstable

Figure 4.6 shows similar pictures to those in Fig. 4.5, but for a block copolymer melt given by (4.35). In this block copolymer case, the Fourier mode with nonzero wave number q^* becomes unstable at about $\chi N \sim 10.5$. This is in contrast to the case of the polymer blend where the Fourier mode with $q = 0$ is destabilized first. This difference comes from the fact that the block copolymer cannot undergo a macrophase separation because the two subchains are connected by a chemical bond. Therefore, the fluctuation with $q = 0$ is inhibited. Instead, the block copolymer undergoes a microphase separation where the two blocks separate with a finite wavelength $2\pi/q^*$ of the same order as the gyration radius of the chain. This situation is just the same as the one discussed in Sect. 3.3.4. The only difference is that the target of the Ginzburg–Landau theory is the weak segregation regime, while the system we discussed in Sect. 3.3.4 is assumed to be in the strong segregation regime. Such a difference leads to a different dependence of the period D^* of the microphase-separated domains on the chain length, i.e., $D^* \sim N^{2/3}$ in the strong segregation regime and $D^* \sim N^{1/2}$ in the weak segregation regime.

When we expand the right-hand side of (4.35) as a power series in q, we obtain the form [34]

$$\frac{1}{\tilde{S}^{(\mathrm{u})}(q)} = \frac{A}{q^2} + B + Cq^2 + \cdots , \tag{4.37}$$

where A, B and C are constants. The first, second and third terms correspond to the elastic energy of the chain, the segment interaction, and the square gradient energy, respectively. The most important term in (4.37) is the first term, which is proportional to $1/q^2$. This accounts for the fact that the two subchains are connected by a chemical bond. Substituting (4.37) into (4.15), the $1/q^2$ term produces the Coulomb-type interaction energy

$$\text{const.} \times \int \mathrm{d}\boldsymbol{r} \int \mathrm{d}\boldsymbol{r}' \, \frac{\delta\phi(\boldsymbol{r})\delta\phi(\boldsymbol{r}')}{|\boldsymbol{r}-\boldsymbol{r}'|} \,, \tag{4.38}$$

where $\delta\phi(\boldsymbol{r}) \equiv \delta\phi_A(\boldsymbol{r}) - \delta\phi_B(\boldsymbol{r})$. There is therefore a similarity between the block copolymer melt system and the system with Coulomb interaction.

Basic Properties of the Ginzburg–Landau Model

As a simplest example of the Ginzburg–Landau model, we consider a symmetric A/B homopolymer blend with the same chain lengths and the same volume fractions of the A and B polymers. Because of the incompressibility condition, the two segment density fields $\delta\phi_A(\boldsymbol{r})$ and $\delta\phi_B(\boldsymbol{r})$ are not independent of each other. We thus introduce a field $X(\boldsymbol{r}) \equiv \delta\phi_A(\boldsymbol{r}) - \delta\phi_B(\boldsymbol{r})$ and expand the free energy with respect to this $X(\boldsymbol{r})$. Truncating the expansion at fourth order, we obtain

$$F = \int \mathrm{d}\boldsymbol{r} \left(-\frac{1}{2}cX^2 + \frac{1}{4}uX^4 + \frac{1}{2}D|\nabla X|^2 \right) . \tag{4.39}$$

Some important properties of the equilibrium states of this free energy, i.e., the solution of $\delta F/\delta X(x) = 0$, are listed below:

- Equilibrium uniform states:

$$\begin{aligned} &X(x) = \pm\sqrt{c/u} \equiv \pm X_{\mathrm{e}} \qquad \text{(stable states)} \,, \\ &X(x) = 0 \qquad \text{(unstable state)} \,. \end{aligned} \tag{4.40}$$

- Inhomogeneous state containing a single interface:

$$X_{\text{interface}}(x) = X_{\mathrm{e}} \tanh\left[(x - x_0)/\xi\right] \,,$$

  where $\xi = \sqrt{2D/c}$ is the interfacial thickness and x_0 is the position of the interface.
- Interfacial tension, i.e., the excess free energy accumulated at the interface per unit interfacial area:

$$\sigma = \frac{2c^2}{3u}\sqrt{\frac{2D}{c}} \,.$$

4.2 Applications of the Ginzburg–Landau Theory

The free energy expansion given in (4.15) can be combined with the RPA results of the correlation functions given in (4.30) to construct a useful free energy model for predicting the qualitative properties of various blends of polymers and block copolymers. With such a model, we can calculate the overall behavior of the phase diagram and the structure of the phase-separated domains and interfaces. In this section, we describe several examples of such studies on block copolymer melts.

The Ginzburg–Landau theory can be applied to basically the same systems that can be treated with the self-consistent field theory. However, the approximate treatment of the free energy in the Ginzburg–Landau theory leads to the following advantages and disadvantages compared to the self-consistent field theory:

- As an explicit expression for the free energy in terms of the segment density is obtained, calculation of the phase diagram and the phase-separated structures is easy, whereas the self-consistent field theory requires a numerical iteration procedure.
- In the strong segregation regime, the higher order terms of the free energy that are neglected in the Ginzburg–Landau expansion give non-negligible contributions. Then the results may be quantitatively (or sometimes qualitatively) wrong.
- A difference in the chain architectures is not fully reflected in the expansion coefficients. This is because the correlation functions are evaluated in the uniform reference system where chains with different architectures show similar coiled conformations.

4.2.1 Phase Diagram of Block Copolymer Melts

As a simple application of the Ginzburg–Landau theory, we calculate the phase diagram of a block copolymer melt. This is the same problem as was discussed in Sect. 3.3.4 using the self-consistent field theory.

As presented in Fig. 4.6, the RPA calculation shows that a symmetric block copolymer melt has a critical point at $\chi N \approx 10.5$. In the vicinity of this critical point, the scattering function is sharply peaked around q^* because this is the only unstable mode at the critical point (see Fig. 4.6b). Thus, the sum over the wave vectors in (4.15) can be approximated by the terms with $|\boldsymbol{q}| = q^*$. To construct the phase diagram of the microphase separation of the block copolymer, we have to retain terms up to fourth order in the Ginzburg–Landau expansion in (4.15). (Recall the behavior of the free energy shown in Fig. 4.2. In order to reproduce such a functional form, at least fourth order polynomial function is necessary.) The coefficients of the terms in the expansion are obtained using the RPA introduced in Sect. 4.1.3.

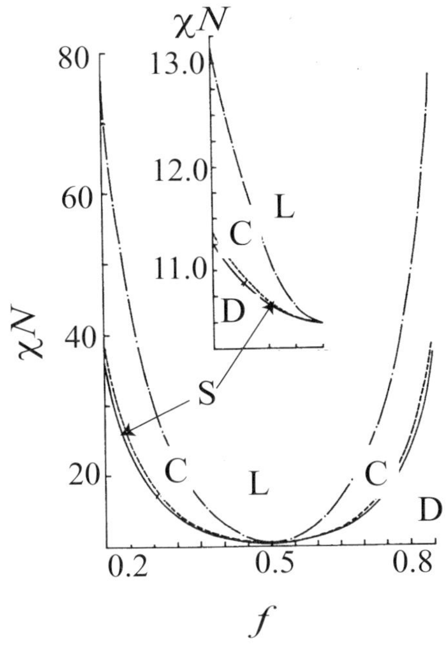

Fig. 4.7. Phase diagram of an AB diblock copolymer melt near its critical point, obtained using the Ginzburg–Landau free energy with the RPA. The *ordinate* is χN and the *abscissa* is the block ratio f. The *symbols* mean L (lamellar), C (cylindrical), S (spherical), and D (disorder). Reprinted with permission from L. Leibler, Macromolecules **13**, 1602 (1980). (© 1980, American Chemical Society)

The procedure for obtaining the phase diagram is almost the same as in the self-consistent field theory. We first list all the candidates for the crystal symmetry of the equilibrium structure. For each of these candidates, we evaluate the Ginzburg–Landau free energy by changing the periodicity of the crystal structure. The equilibrium structure and its periodicity are the ones that give the global minimum of the free energy. The phase diagram obtained with the Ginzburg–Landau model is shown in Fig. 4.7 [39]. This phase diagram corresponds to the one presented in Fig. 3.29, which was obtained with the self-consistent field theory. We can confirm that the behavior of these two phase diagrams around the critical point coincides well. This is because the Ginzburg–Landau theory and the self-consistent field theory rely on the same assumptions (the mean field approximation, a reference state composed of independent ideal chains, etc.). On the other hand, the agreement between the two phase diagrams becomes worse in the strong segregation region because of the large truncation error in the Ginzburg–Landau free energy.

As discussed in this section, the Ginzburg–Landau theory and the self-consistent field theory are complementary. The Ginzburg–Landau theory is useful for studying the qualitative behavior of the system in the weak segregation regime. However, for quantitative discussions, especially in the strong segregation regime, one should use the self-consistent field theory.

Although the power series expansion of the free energy seems to be valid in the weak segregation regime, there is another complication. As the system approaches the critical point, the thermal fluctuations tend to be more and more important. However, the mean field approximation used in the derivation of the Ginzburg–Landau free energy and its coefficients neglects the effects of thermal fluctuations. Thus the phase diagram shown in Fig. 4.7 may be unreliable. To overcome this difficulty, one has to explicitly include the effects of thermal fluctuations. This can be done using the renormalization group method [40].

As discussed briefly in Sect. 2.3.3, the renormalization transformation traces the change in the parameters of the free energy model when the model is coarse-grained. During this coarse-graining process, the effects of thermal fluctuations are taken into account in a step-by-step manner. The final form of the free energy after the renormalization transformation is usually different from the one obtained by the mean field theory. In the block copolymer melt case, it was shown that this renormalization transformation produces extra terms that are not included in the mean field free energy, and this drastically changes the nature of the phase transition at the critical point. Figure 4.8 shows the phase diagram of a block copolymer melt similar to Fig. 4.7, but including thermal fluctuation effects. In this phase diagram, one can see that the phase boundaries around the critical point are modified. Moreover, the second order transition around the critical point of the mean field model turns into a first order phase transition in this calculation due to the renormalization effect. Such a first order phase transition of block copolymer melts at the critical point is confirmed experimentally.

4.2.2 Extensions to Dynamical Processes

Similarly to the dynamical version of the self-consistent field theory presented in Sect. 3.3.3, we can construct a dynamical model of polymer systems based on the Ginzburg–Landau model or the Flory–Huggins–de Gennes model. Such a dynamic version of the Ginzburg–Landau model is often referred to as the time-dependent Ginzburg–Landau model and abbreviated to the TDGL model in the literature.

When simulating time evolution problems such as the dynamics of phase separation, the main problem is the need for a large simulation box. Because the domains in phase-separating systems grow continuously, the size of the domains finally reaches the size of the simulation box, at which stage the simulation results are significantly affected by the boundary condition of the simulation box. Thanks to the computational efficiency of the Ginzburg–Landau model, it is more suitable for large-scale dynamical simulations than the dynamic self-consistent field method as long as we are only interested in the qualitative behavior of the system.

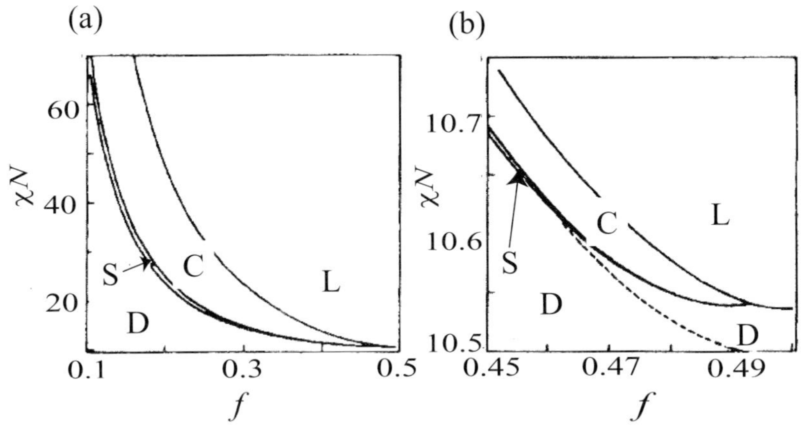

Fig. 4.8. Phase diagrams of a symmetric AB diblock copolymer melt calculated using the Ginzburg–Landau free energy. The *ordinate* is χN and the *abscissa* is the block ratio $f = N_A/(N_A + N_B)$. (**a**) The case where the effects of the thermal fluctuations are neglected (the same as given in Fig. 4.7) and (**b**) the case where the effects of the thermal fluctuations are taken into account. *Symbols* show L (lamellar), C (cylindrical), S (spherical) and D (disorder) phases. Reprinted with permission from G.H. Fredrickson and E. Helfand, J. Chem. Phys. **87**, 697 (1987). (© 1987, American Institute of Physics)

In order to check the validity of the time-dependent Ginzburg–Landau model, we show in Fig. 4.9 a comparison between the results of the time-dependent Ginzburg–Landau model based on the Flory–Huggins–de Gennes free energy and the results of the dynamic self-consistent field model for a symmetric A/B polymer blend containing a symmetric or asymmetric AB block copolymer as compatibilizer. It has been confirmed that these two dynamical models give qualitatively the same results, not only for the morphology of the domains, but also for the temporal growth of the average domain size [41], as long as the parameters of the time-dependent Ginzburg–Landau model are chosen appropriately.

Another suitable target for the time-dependent Ginzburg–Landau model is the dynamics of the structural phase transition between different types of microphase-separated phase. This problem is the dynamical extension of the problem discussed in Fig. 3.32 using the self-consistent field theory. Figure 4.10 shows results of a simulation using the time-dependent Ginzburg–Landau model on the time evolution of the domain structure from an initial cylindrical one to a final spherical one [42]. We observe the onset of undulations in the cylindrical domain after the sudden temperature quench (Fig. 3.32b). A detailed analysis on the behavior of the free energy during this transition process reveals that this undulating cylindrical structure corresponds to a saddle point of the free energy. Thus, the rate of the structural

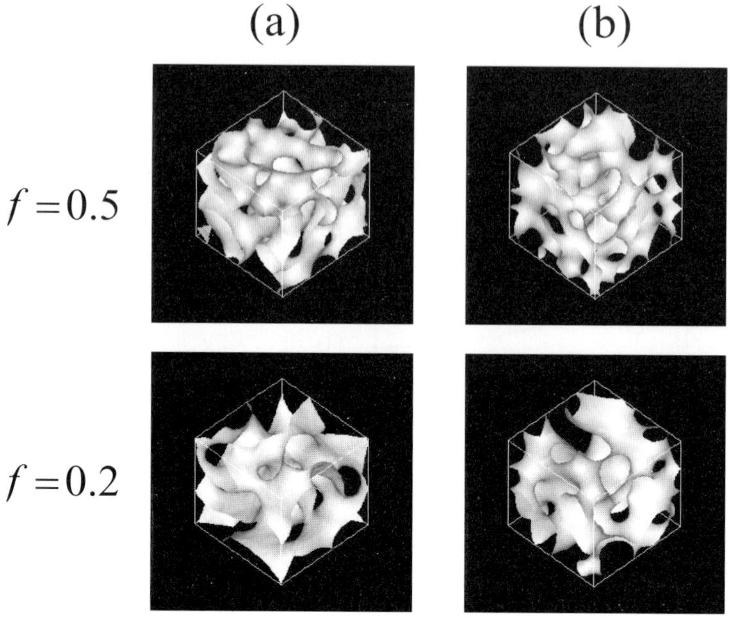

Fig. 4.9. Domain structures in a symmetric A/B homopolymer blend containing a symmetric ($f = 0.5$) or asymmetric ($f = 0.2$) AB diblock copolymer as compatibilizer. Each domain structure is obtained by a dynamical simulation using (**a**) the self-consistent field theory or (**b**) the Flory–Huggins–de Gennes model. (In the latter case, the coefficient of the square gradient term should be calculated in a similar way to the polymer blend case described in the text)

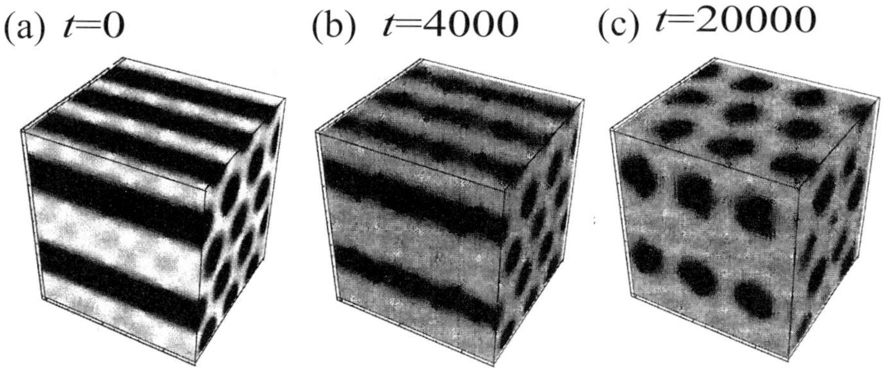

Fig. 4.10. Structural phase transition of an AB block copolymer melt from the cylindrical phase to the spherical phase. The simulation has been carried out using the time-dependent Ginzburg–Landau model. (**a**) Snapshots just after the temperature quench, and (**b**), (**c**) at later times. Reprinted with permission from S. Qi and Z.-G. Wang, Phys. Rev. E **55**, 1682 (1997). (© 1997, American Physical Society)

transition is dominated by the rate of change across this saddle point (i.e., this is the so-called rate-determining process).

In order to understand the overall dynamical behavior of the structural phase transition, it is important to obtain the so-called free energy landscape, i.e., the free energy surface as a function of the control parameters (e.g., temperature) and the state of the system (e.g., the domain morphology). Because the calculation of such a free energy landscape using the self-consistent field theory usually requires considerable computer power, the Ginzburg–Landau model is useful for understanding the qualitative dynamics of the structural phase transition.

Exercises

Exercise 4.1. Expand the scattering functions for the homopolymer blend (4.32) and for the block copolymer melt (4.35) in power series in the wave number $|\boldsymbol{q}|$. Then obtain the coefficient of the square gradient term in the free energy $\nabla \delta \widetilde{\phi}_K(\boldsymbol{q})$. (The terminology 'Ginzburg–Landau model' usually refers to the free energy after this expansion with respect to $|\boldsymbol{q}|$ has been performed.)

Exercise 4.2. By applying the random phase approximation to a melt of a star block copolymer or a melt of a comb block copolymer, calculate the scattering function for these systems.

5 Macroscopic Viscoelastic Theory of Polymers

The aim of this chapter is to discuss the macroscopic hydrodynamic properties of polymeric liquids and relate them to microscopic chain dynamics. Polymeric liquids possess characteristic features of both viscous fluids and elastic bodies. Such dynamic behavior is called viscoelasticity. The origin of the viscoelasticity of polymeric liquids is the complex internal structure of the polymer chains and mutual entanglements between the chains. In this chapter, we first formulate viscoelasticity on the macroscopic scale using hydrodynamic descriptions. We then introduce reptation theory, which is a coarse-grained molecular theory of viscoelastic behavior in dense entangled polymer liquids.

5.1 Viscoelastic Properties of Polymeric Liquids

5.1.1 Polymers and Viscoelasticity

Viscous Liquids and Elastic Bodies

So far we have been discussing polymeric systems focusing mainly on equilibrium systems or weakly non-equilibrium systems close to equilibrium. In this chapter, we study the dynamical properties of highly non-equilibrium polymeric systems, such as a chain (or a network of chains) highly deformed by an external force field or an external flow field. A remarkable property of polymeric liquids which distinguishes them from systems composed of small molecules is viscoelasticity. A viscoelastic body, i.e., a material that exhibits viscoelasticity, possesses features of both viscous fluids and elastic bodies.

When we stir a low-molecular-weight liquid such as water or silicon oil, the induced flow will gradually decay and will finally vanish unless we continue to stir. The origin of the decay of the flow is the friction arising from the inhomogeneity in the flow field. As shown in Fig. 5.1, let us consider a virtual plane perpendicular to the direction of the velocity gradient. Due to the velocity gradient between both sides of the plane, a friction force is generated in such a direction that the velocity difference is reduced. Such a friction force is called the shear stress. A viscous fluid is defined as a fluid that generates shear stress when there is inhomogeneity in the flow velocity.

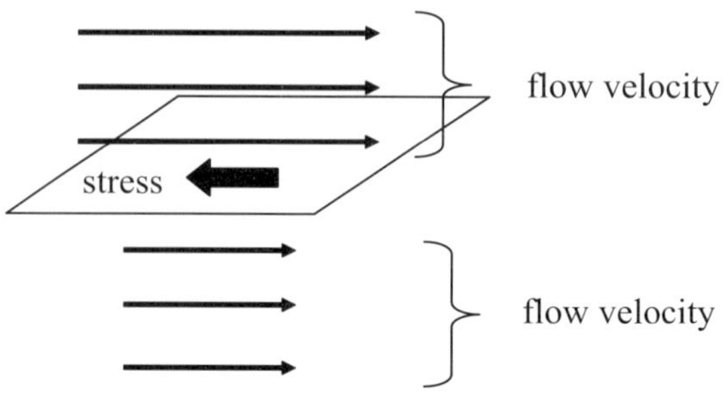

Fig. 5.1. Intuitive illustration of the shear stress. If we imagine a virtual plane that is perpendicular to the velocity gradient, there is a difference between the fluid velocities on either side of this plane. Then the slower fluid pulls the faster fluid, and vice versa. This effect leads to a friction force that diminishes the inhomogeneity in the flow velocity

If the shear stress is proportional to the spatial gradient of the flow velocity, such a viscous fluid is called a linear viscous fluid or a Newtonian fluid. The characteristic property of the viscous fluids is that the shear stress depends only on the inhomogeneity in the instantaneous flow velocity. Thus, a viscous fluid completely loses its memory on the imposed deformation or on the imposed flow after a certain relaxation time. This is why a viscous fluid can fill a container with an arbitrary shape.

On the other hand, a rubber produces a restoring force when it is externally deformed. This restoring force will not vanish unless the external deformation is removed. Such a material is called an elastic body. The measure of the deformation of an elastic body is the strain, which is defined as the spatial gradient of the displacement of each material point. (The precise mathematical definition of the strain will be given later.) Unlike the viscous fluid, an elastic body can produce a shear stress even if it is in a static state without a flow. When the shear stress is proportional to the strain, it is called a linear elastic body or a Hookian elastic body.

Dynamic Properties of Viscoelastic Bodies

A viscoelastic body behaves as an elastic body against a fast deformation and as a viscous fluid against a slow deformation. One such example is a toy named 'slime', which is a rubber-like viscous fluid. The slime is made of a dense aqueous solution of a polymer named polyvinyl alcohol, where the chains are crosslinked by weak hydrogen bonds to form a temporary network (see Fig. 5.2).

When we flick the slime quickly with a finger, it vibrates just like an elastic body. On the other hand, if we put the slime on a slope, it flows down like a viscous fluid. The main difference between these two types of behavior is the difference in the characteristic time scales of the externally imposed forces.

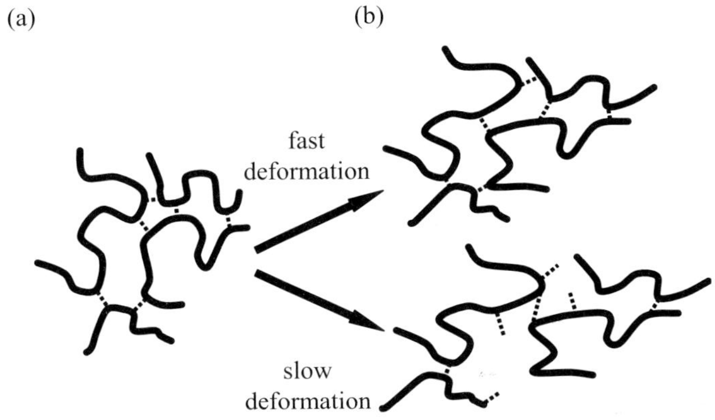

Fig. 5.2. (**a**) Structure of slime on the molecular scale. Polyvinyl alcohol chains are crosslinked by hydrogen bonds (*broken lines*) and form a network structure. (**b**) Response of the network to fast and slow deformations. When the deformation is slow, the hydrogen bonds are broken and then reconnect

While the characteristic time scale of the impulse when we flick the slime with a finger is less than about 1 s, the characteristic time scale of the flow of the slime on a slope due to the gravity is of the order of 1 min. Therefore, we recognize that the slime changes its response at a certain crossover time scale around 1 s.

Using a molecular-scale description, this complex behavior is easily understood. Under a fast deformation, the network formed by the hydrogen bonds deforms quickly without reconnecting, which leads to an elastic response. When a slow deformation is imposed, on the other hand, the hydrogen bonds are gradually broken, and reconnections of the network take place. This leads to the fluid-like behavior. Therefore, the deformation and reconnection of the temporary network formed by the hydrogen bonds lie at the origin of the viscoelastic behavior of the slime. Similar viscoelastic behavior is observed in dense polymer solutions or in polymer melts, even if they do not have the hydrogen bonds. In such systems, the origin of the viscoelasticity is the temporary network formed by dynamic entanglements between polymer chains.

Die Swell

One of the characteristic properties of viscoelastic fluids is the finite memory of the history of the flow or the deformation in the past. Such a memory is stored in the material in the form of the deformation of the network. A typical example of this memory effect is the phenomenon known as die swell [8]. When a viscous fluid is extruded from a die (a narrow channel), its diameter reduces because the flow velocity increases due to the acceleration by gravity (see Fig. 5.3a). In the case of a viscoelastic fluid, the behavior is opposite. Its diameter increases when it is extruded from a die (see Fig. 5.3b). Such anomalous behavior of viscoelastic fluids is called die swell. The origin of this phenomenon is the recoil of the deformed molecular network. Because the molecular network is highly deformed by the shear deformation inside the narrow channel, it recoils when it gets out of the channel and the shear strain is suddenly removed.

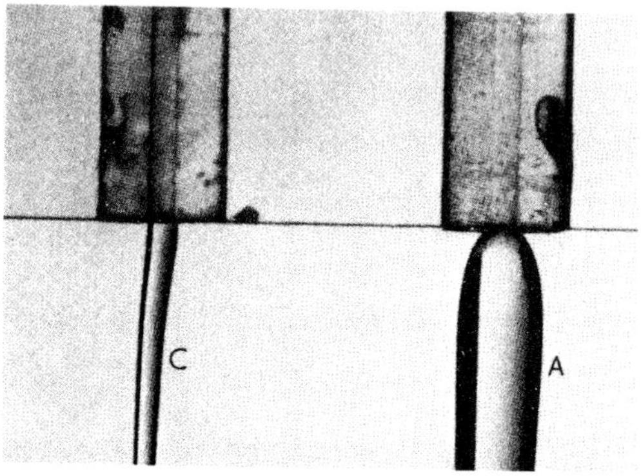

Fig. 5.3. Change in the diameter of a fluid flowing out of a narrow channel (die). Comparison between (**a**) silicon oil (a viscous fluid) and (**b**) a polymethyl methacrylate aqueous solution (a viscoelastic fluid). Reprinted with permission from A.S. Lodge, *Elastic Liquids* (© Academic Press, London 1964)

Storage and Loss Moduli

A quantitative measure of the viscoelasticity is the elastic modulus, defined as the proportionality constant between the imposed strain and the generated stress due to the strain. Experimental data concerning the elastic moduli are obtained by measuring the stress of the sample when various deformations are imposed on it. One such technique is the stress relaxation measurement, in which an instantaneous step deformation is imposed on the sample and the stress relaxation after this sudden deformation is measured. Another technique is the dynamic elasticity measurement, in which the stress is measured while a small oscillatory deformation is imposed steadily on the sample. (Instead of imposing a deformation, one can also impose a stress on the sample. The creep measurement is one such example, in which the time dependence of the strain of the sample is measured under a constant shear stress starting at a certain time.) For the simple deformations shown in Fig. 5.4, the strain is characterized by the scalar variable γ defined by[1]

[1] In general, the strain is described by a tensor.

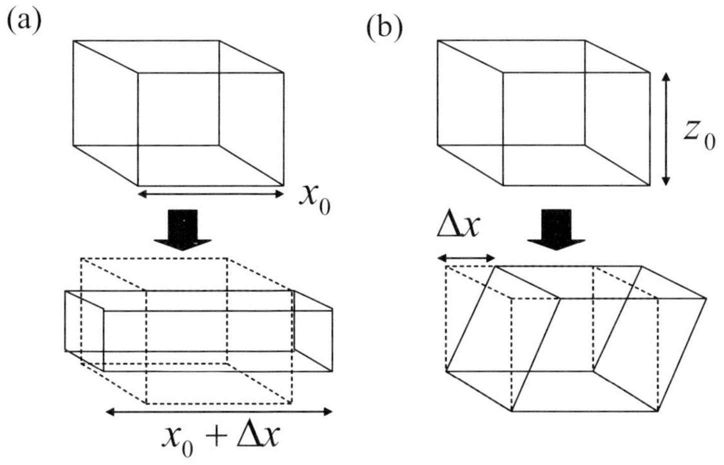

Fig. 5.4. Examples of simple deformations. (**a**) A uniaxial extension and (**b**) a simple shear deformation

$$\gamma = \begin{cases} \Delta x / x_0 & \text{(uniaxial deformation)}\ , \\ \Delta x / z_0 & \text{(shear deformation)}\ . \end{cases} \tag{5.1}$$

Using this definition of the strain, we will explain how to perform the dynamic elasticity measurement. In order to measure the shear stress $\sigma(t)$, let us impose the following oscillatory strain on the specimen:

$$\gamma(t) = \gamma_0 \sin \omega t\ , \tag{5.2}$$

where γ_0 is the maximum amplitude of the oscillatory strain (γ_0 is assumed to be time-independent) and ω is the angular frequency of the oscillation (see Fig. 5.5).

The shear stress oscillates with the same frequency as the imposed strain $\gamma(t)$, but in general with a phase shift:

$$\sigma(t) = \sigma_0 \sin(\omega t + \delta)\ . \tag{5.3}$$

If the specimen is an elastic body, the stress is determined by the instantaneous strain, and there is no phase shift, i.e., $\delta = 0$. On the other hand, for a viscous fluid, the stress is determined by the gradient of the flow velocity. Note that the flow velocity is the time derivative of the displacement of each point fixed to the fluid. As the spatial gradient of such a displacement gives the strain γ, the gradient of the flow velocity is given by $\dot{\gamma}(t)$, where the dot denotes the time derivative (see Fig. 5.5). Therefore, for a viscous fluid, the phase shift δ between the strain and the stress is given by $\delta = \pi/2$.

For a general viscoelastic body, its behavior lies somewhat in-between the above two extreme cases. In this case, the value of the phase shift δ differs from material to material, and it also depends on the frequency ω.

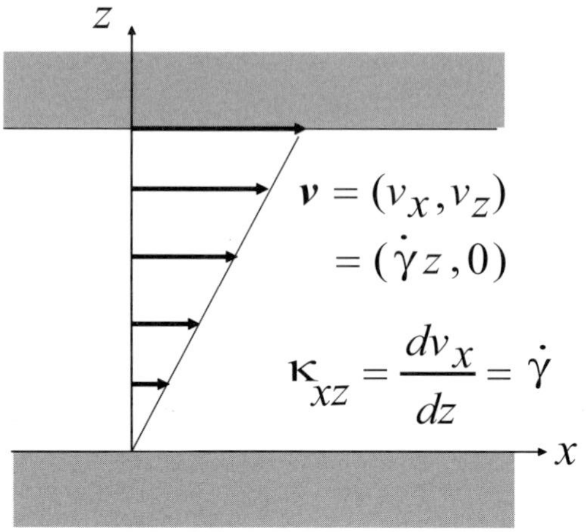

Fig. 5.5. Illustration of a shear flow and the xz-component of the velocity gradient tensor. The definition of the velocity gradient tensor is given in (5.9)

The quantity δ is thus a useful measure of the viscoelasticity. To characterize the viscoelasticity using δ, let us introduce the following two quantities $G'(\omega)$ and $G''(\omega)$:

$$G'(\omega) = \frac{\sigma_0}{\gamma_0} \cos\delta(\omega) \, , \qquad G''(\omega) = \frac{\sigma_0}{\gamma_0} \sin\delta(\omega) \, . \tag{5.4}$$

Substituting the two equations (5.4) into (5.3), we obtain the relation

$$\begin{aligned} \sigma(t) &= \gamma_0 \Big[G'(\omega) \sin\omega t + G''(\omega) \cos\omega t \Big] \\ &= G'(\omega)\gamma(t) + \frac{G''(\omega)}{\omega} \dot{\gamma}(t) \, , \end{aligned} \tag{5.5}$$

which clearly shows that $G'(\omega)$ and $G''(\omega)$ correspond to the elastic response and viscous response, respectively. These $G'(\omega)$ and $G''(\omega)$ are called the storage modulus and the loss modulus. As is obvious from the above discussion, the storage modulus G' is large for elastic bodies and the loss modulus G'' is large for viscous fluids. Hence, $G'(\omega)$ and $G''(\omega)$ indicate whether the material behaves as an elastic body or as a viscous fluid on the time scale of $2\pi/\omega$.

Elastic Moduli of Dense Polymer Systems

Figure 5.6 shows the frequency dependence of the storage modulus G' of polystyrene melts obtained by a dynamic viscoelastic measurement [43]. The experimental data show that the polystyrene melt behaves as an elastic body in the high frequency region (short time scale), and as a viscous fluid in the

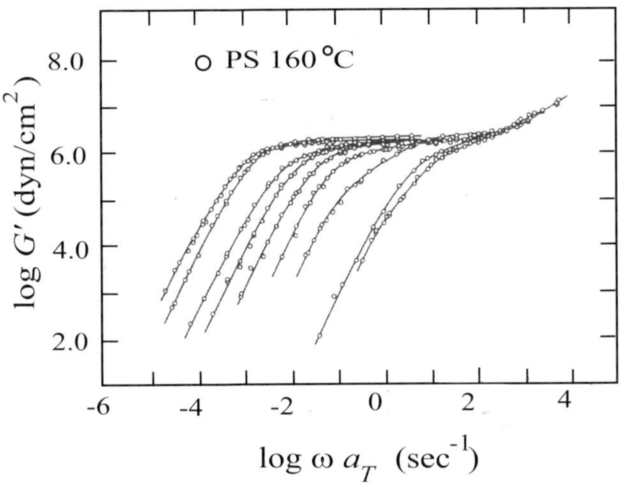

Fig. 5.6. Experimental data on the storage modulus $G'(\omega)$ of polystyrene melts with various polymerization indices. *From right to left*, the polymerization index increases. Reprinted with permission from S. Onogi et al., Macromolecules **3**, 109 (1970). (© 1970, American Chemical Society)

low frequency region (long time scale). For polymers with high molecular weights, there is a plateau in the intermediate frequency region called the rubber-elasticity plateau. Within this plateau regime, the polymer melt behaves just like an elastic body with a constant elastic modulus independent of the frequency.

The main aim of this chapter is to derive macroscopic viscoelastic properties, such as the frequency-dependent elastic moduli $G'(\omega)$ and $G''(\omega)$, from the microscopic molecular models of polymer chains discussed in Chap. 2. The microscopic origin of viscoelasticity is very different in dilute polymer solutions and in dense polymer solutions. In a dilute polymer solution, the viscoelastic properties arise from the conformational relaxation of isolated chains that are deformed by the external force. In this case, the viscoelastic properties can be derived from the relaxation mechanism of the Rouse or Zimm models (see Sects. 2.4.2 and 2.4.3).

The viscoelasticity of dense polymer systems is much more complex. Microscopically, a deformation of a dense polymer system means a deformation of the temporary network formed by the entangling polymer chains. The important difference that distinguishes polymer melts from rubbers is the fact that such a deformed network relaxes due to the disentanglement of polymers. This is the origin of the long-time stress relaxation of polymer melts (see Fig. 5.7). For a theoretical analysis of the viscoelastic properties of dense polymer systems, we have to consider the motion of the individual chain in

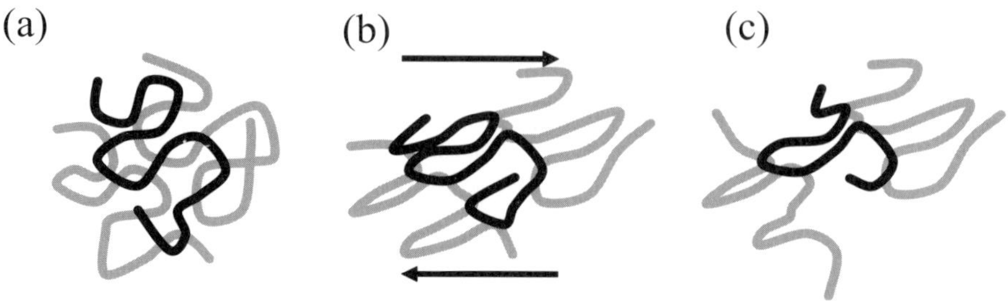

Fig. 5.7. The concept of viscoelasticity in a dense polymer system. (**a**) Network formed by mutually entangling chains. (**b**) Deformation of the network due to an external force. (**c**) Relaxation of a chain in the network

a network composed of many entangling chains (Fig. 5.7c). Reptation theory is one successful theory for this purpose. We will discuss it in Sect. 5.2.

5.1.2 Hydrodynamic Descriptions of Viscoelasticity

Basic Equations for Viscoelasticity

This section is devoted to a brief summary of the basic equations of viscoelasticity, which will be used in the following sections. For details, readers should refer to standard textbooks [6, 8].

Our starting point is the equation of continuity and the equation of motion of the fluid. Because dense polymer systems are generally highly incompressible and highly viscous, we neglect both the compressibility and the inertia. (The latter corresponds to the Stokesian approximation.) Within this approximation, the equation of continuity and the equation of motion are written as

$$\nabla\cdot\boldsymbol{v} = 0 \tag{5.6}$$

and

$$\rho\frac{\partial \boldsymbol{v}}{\partial t} = \nabla\cdot\mathit{\Pi} + \rho\boldsymbol{f} \ , \tag{5.7}$$

where ρ is the total density of the fluid, which is kept constant, $\boldsymbol{v}(\boldsymbol{r},t)$ is the flow velocity at time t at position $\boldsymbol{r}$, and $\mathit{\Pi}(\boldsymbol{r},t)$ and $\boldsymbol{f}(\boldsymbol{r},t)$ are the stress tensor and the body force acting on the fluid, respectively. Examples of the body force $\boldsymbol{f}$ are gravity, the thermodynamic force induced by interfaces, and so on. Then the stress tensor is a sum of the isotropic hydrostatic pressure and the shear viscosity,[2] with the general form

$$\boldsymbol{\mathit{\Pi}}(\boldsymbol{r},t) = -p(\boldsymbol{r},t)\mathbf{1} + \boldsymbol{\sigma}(\boldsymbol{r},t) \ . \tag{5.8}$$

[2] There is another contribution to the viscosity, which is called the bulk viscosity. As this bulk viscosity is associated with the volume change of the fluid, it drops out for incompressible fluids.

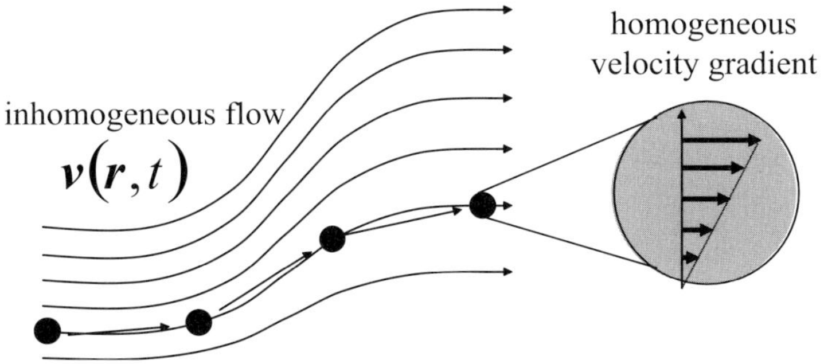

Fig. 5.8. The concept of a fluid particle. A fluid particle (*solid circle*) moves with the same velocity as the fluid flow along the stream path

The first term on the right-hand side is the hydrostatic pressure, where p is the magnitude of the pressure and $\mathbf{1}$ is the unit tensor. The second term accounts for the stress caused by the deformation. (If the inertia cannot be neglected, the stress tensor is replaced by the momentum flux density tensor, which includes the contribution from advection. For viscous polymeric fluids, such a contribution can usually be neglected.) The characteristic viscoelastic properties of polymers come from the dependence of $\boldsymbol{\sigma}(\boldsymbol{r}, t)$ on the history of the deformation.

The history of the imposed deformation on the fluid is memorized in the form of a change in the internal state of the fluid, such as the chain conformation or the domain shape. It is therefore important to observe the deformation in a frame moving with the fluid flow. A useful concept for this purpose is the fluid particle illustrated in Fig. 5.8. A fluid particle is defined as an infinitesimal volume element that moves with the fluid flow.

We assume that the fluid particle is large enough to be able to apply the hydrodynamic description, but that it is still small to justify neglecting any inhomogeneity inside it. With these assumptions, it is enough to discuss only the spatially uniform deformations.

Strain and Velocity Gradient Tensor

The deformation of a fluid can be described in terms of the gradient of the flow velocity. This quantity is a second-rank tensor composed of 9 elements, i.e., combinations of the three components of spatial gradient operators, $\partial/\partial x$, $\partial/\partial y$ and $\partial/\partial z$, and the three components of the flow velocity, v_x, v_y and v_z. This tensor is called the velocity gradient tensor and is defined by

$$\boldsymbol{\kappa} = (\nabla \boldsymbol{v})^{\mathrm{T}} , \tag{5.9}$$

where $\nabla \boldsymbol{v}$ is a tensor whose $\alpha\beta$ element ($\alpha, \beta = x, y, z$) is given by $(\partial/\partial_\alpha)v_\beta$ and the superscript T indicates the transpose of the matrix. As shown in

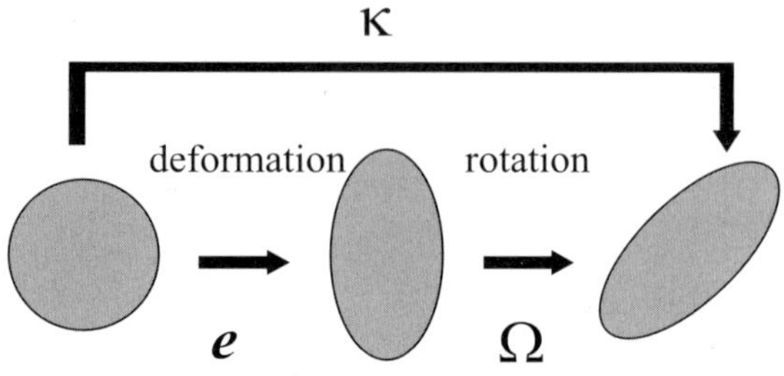

Fig. 5.9. The displacement of a fluid can be decomposed into a deformation and a rotation

Fig. 5.9, the velocity gradient tensor is in general composed of two contributions, a deformation and a rotation.

This is easily proved by rewriting (5.9) in the form

$$\boldsymbol{\kappa} = \frac{1}{2}\left[(\nabla \boldsymbol{v})^{\mathrm{T}} + \nabla \boldsymbol{v}\right] + \frac{1}{2}\left[(\nabla \boldsymbol{v})^{\mathrm{T}} - \nabla \boldsymbol{v}\right] \equiv \boldsymbol{e} + \boldsymbol{\Omega}\,. \tag{5.10}$$

The first and second terms on the right-hand side are the symmetric and antisymmetric parts of the velocity gradient tensor, respectively. The antisymmetric part $\boldsymbol{\Omega}$ can be rewritten in terms of the rotation of the vector $\boldsymbol{v}$, i.e., $\boldsymbol{\omega} = (\omega_x, \omega_y, \omega_z) \equiv \operatorname{rot} \boldsymbol{v}$, as

$$\boldsymbol{\Omega} = \frac{1}{2}\begin{pmatrix} 0 & -\omega_z & \omega_y \\ \omega_z & 0 & -\omega_x \\ -\omega_y & \omega_x & 0 \end{pmatrix}. \tag{5.11}$$

Thus, any displacement vector $\boldsymbol{\delta r}$ satisfies the relation

$$\boldsymbol{\Omega} \cdot \boldsymbol{\delta r} = \frac{1}{2}\boldsymbol{\omega} \times \boldsymbol{\delta r}\,, \tag{5.12}$$

which means that $\boldsymbol{\Omega}$ is a simple rotation and will not cause any deformation. Thus, only the first term on the right-hand side of (5.10), i.e., $\boldsymbol{e}$, causes stress. This tensor $\boldsymbol{e}$ is called the deformation velocity tensor.

Constitutive Equation of Viscoelasticity

Using the velocity gradient tensor or the deformation velocity tensor, the stress of the fluid is written as a functional of the history of the velocity gradient as

$$\boldsymbol{\sigma}(\boldsymbol{r}, t) = \boldsymbol{S}\left[\{\boldsymbol{\kappa}(\boldsymbol{r}, t'); t' < t\}\right], \tag{5.13}$$

where $\boldsymbol{\kappa}$ on the right-hand side refers to the history of the deformation of the fluid particle along the stream path, which introduces a non-local effect on the right-hand side. Equation (5.13) is called the constitutive equation.

If the imposed strain is small, the stress–strain relation is well approximated by a linear relation. For such a system, the constitutive equation is written as [6,8]

$$\boldsymbol{\sigma}(t) = \int_{-\infty}^{t} \mathrm{d}t' G(t-t')\left[\boldsymbol{\kappa}(t') + \boldsymbol{\kappa}^{\mathrm{T}}(t')\right] , \tag{5.14}$$

where we have used the fact that only the deformation velocity tensor generates the stress. A viscoelastic body whose constitutive equation is given by the linear equation (5.14) is called a linear viscoelastic body.

Let us consider the response of a linear viscoelastic body to a step strain, where a strain $\boldsymbol{\gamma}_0$ is suddenly imposed at $t = 0$ and kept constant afterwards. In this case, the velocity gradient tensor $\boldsymbol{\kappa}$, which is the time derivative of the strain tensor, leads to

$$\boldsymbol{\kappa}(t) = \boldsymbol{\gamma}_0 \delta(t) , \tag{5.15}$$

where $\delta(t)$ is the Dirac delta function. For the simple shear deformation shown in Fig. 5.5, $\boldsymbol{\gamma}_0$ is given by

$$\boldsymbol{\gamma}_0 = \begin{pmatrix} 0 & 0 & \gamma_0 \\ 0 & 0 & 0 \\ 0 & 0 & 0 \end{pmatrix} . \tag{5.16}$$

Substituting this into (5.14), one immediately sees that the stress relaxation for a step strain equals the coefficient $G(t)$ in (5.14). Due to this property, $G(t)$ is called the stress relaxation function. It is now an easy exercise to confirm that the real and imaginary parts of $\mathrm{i}\omega\tilde{G}(\omega)$ give $G'(\omega)$ and $G''(\omega)$ in (5.4), respectively, where $\tilde{G}(\omega)$ is the Fourier transformation of $G(t)$ (see Exercise 5.1).

When the imposed strain is so large that the chains deform considerably, the linear relation (5.14) no longer holds. This kind of viscoelasticity is called nonlinear viscoelasticity. Even for such nonlinear viscoelasticity, the quantity that characterizes the viscoelastic behavior is the stress relaxation function $G(t)$, defined as the stress generated by a step strain as in the case of linear viscoelasticity.

Simple Models of Linear Viscoelasticity

The simplest model of linear viscoelasticity is the Maxwell model. This model comprises a serial connection of two elements, i.e., one is a spring with elasticity G and the other is a dashpot with viscosity η, as shown in Fig. 5.10a. The spring responds to rapid deformations elastically while the dashpot responds to slow deformations as a viscous fluid. The Maxwell model thus shows either elasticity or viscosity depending on the characteristic time scale of the deformation. When a strain $\gamma(t)$ is imposed on the Maxwell model, the stress $\sigma(t)$ is given as a superposition of the responses of the two elements as

$$\frac{\mathrm{d}}{\mathrm{d}t}\sigma(t) = -\frac{G}{\eta}\sigma(t) + G\frac{\mathrm{d}\gamma(t)}{\mathrm{d}t} . \tag{5.17}$$

Another simple model of linear viscoelasticity is the Voigt model which comprises a spring and a dashpot connected in parallel as shown in Fig. 5.10b. While the Maxwell model behaves as a viscous fluid under slow deformation, the Voigt model behaves as an elastic body against a slow deformation because of the spring force. The constitutive equation of the Voigt model is given by

$$\sigma(t) = G\gamma(t) + \eta\frac{\mathrm{d}\gamma(t)}{\mathrm{d}t} . \tag{5.18}$$

Although (5.17) and (5.18) appear to be different from the form of (5.14), we can show that they can actually be rewritten in the form (5.14). To confirm this, one should Fourier transform both sides of (5.17) or (5.18) to obtain an equation of the form

$$\tilde{\sigma}(\omega) = \mathrm{i}\omega\tilde{G}(\omega)\tilde{\gamma}(\omega) . \tag{5.19}$$

Then inverse Fourier transformation of this equation gives the constitutive equation in the form (5.14).

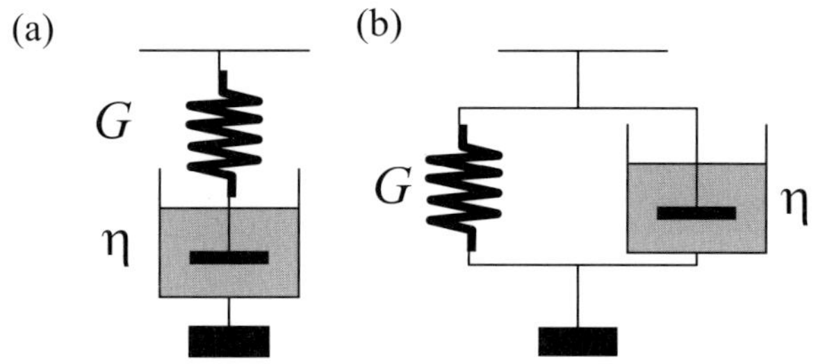

Fig. 5.10. (**a**) Maxwell model and (**b**) Voigt model

5.2 Reptation Theory for Linear Polymers

5.2.1 Concept of Reptation Theory

Tube Model

The reptation theory is one of the standard theories for the linear and nonlinear viscoelasticity of dense polymers based on a coarse-grained molecular picture [6]. In the following, we will discuss both linear and nonlinear viscoelasticity using the reptation theory.

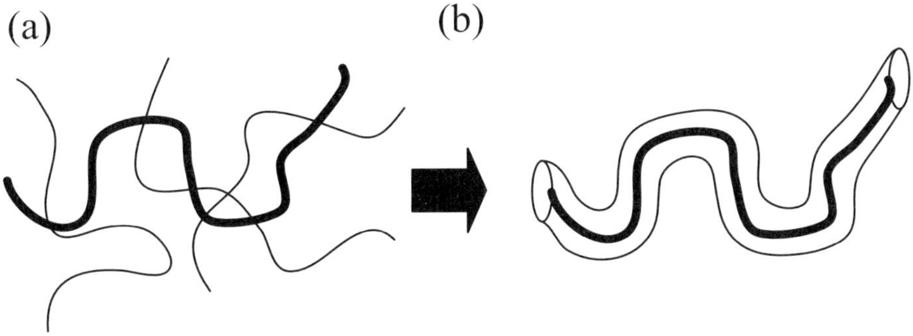

Fig. 5.11. The concept of reptation theory. (**a**) A single chain (tagged chain) in a network formed by entangled chains. (**b**) In the reptation theory, the surrounding chains are replaced by a tube

The central idea of reptation theory is to introduce the concept of a tube. As shown in Fig. 5.7, when the temporary network in a dense polymer system is deformed by an external strain, the deformation of the network gradually relaxes due to disentanglement. The tube picture is used to describe both the entanglement and the disentanglement of polymers quantitatively.

Let us focus on a single chain in the network (hereafter called the tagged chain). As the chains cannot cross each other due to the steric constraint, the tagged chain can move only in the direction parallel to its coarse-grained contour (see Fig. 5.11a).

In the reptation theory, such a steric constraint is approximated by an effective 'tube'. As the many-body effects are replaced by a single tube, this modeling is a kind of mean field approximation. The tagged chain is modeled as a one-dimensional Rouse chain (see Sect. 2.4.2) confined in the tube, and the motion of the segments of the tagged chain is restricted to the direction parallel to the tube.

Diameter and Equilibrium Length of Tube

Let us denote the segment size and the number of segments of the Rouse chain confined in the tube by b and N, respectively. As the tube accounts for the entanglements, the diameter of the tube corresponds to the average distance between nearest entanglement points, i.e., the average size of the mesh formed by the polymer network. Let us denote this length by a. In Sect. 2.3.4, we discussed the screening effect of dense polymer systems, i.e., the excluded volume interaction between segments is cancelled due to the incompressibility condition. In this case, the probability distribution of the chain conformation obeys ideal chain statistics. Let us define a coarse-grained segment whose diameter is the same as the tube diameter a. The scale-free nature of the ideal chain (see Sect. 2.3) guarantees that the conformational statistics of the coarse-grained chain are still the ideal chain statistics. Denoting the total number of segments contained in a coarse-grained segment by N_e, we obtain the following relation using the scaling nature of the end-to-end distance of an ideal chain:

$$a \sim N_e^{1/2} b \,. \tag{5.20}$$

Hereafter, we shall refer to this N_e as the number of segments between entanglement points. (Instead of this N_e, the molecular weight between entanglement points M_e is usually used. As the difference between them is only a constant factor, the results obtained in this chapter remain intact when we use M_e instead of N_e.) Using this N_e, the total number Z of coarse-grained segments in a chain is given by

$$Z = \frac{N}{N_e} \,. \tag{5.21}$$

In this picture, the tube is described as a sequence of Z short cylindrical elements enclosing the chain, where both the diameter and the length of the individual cylindrical element are equal to a. Using the tube diameter a as the unit of length, we can measure the total length of the tube l_{eq}. As each of the Z cylindrical elements contributes a to the tube length, the total length of the tube is given by

$$l_{eq} = aZ = a\frac{N}{N_e} = \frac{b^2 N}{a} \,, \tag{5.22}$$

where we have used the relations (5.20) and (5.21) in the second and third equalities. Equation (5.22) is equivalent to the relation $a^2 Z = b^2 N$. This equation gives two different ways of measuring the end-to-end distance of the chain using two different units of length, a and b.

5.2.2 Stress Relaxation Function

Bond Distribution and Stress Tensor

Let us calculate the stress relaxation function $G(t)$ of a polymeric system based on the reptation model. First, we derive an expression for the stress tensor based on the bead–spring model without the excluded volume interaction, whose equilibrium state is equivalent to that of the Rouse model. Figure 5.12a shows the definition of the stress tensor $\boldsymbol{\sigma}$. Stress is a surface force acting on a virtual plane through which the material on one side of this plane imposes a force on the material on the other side. The $\alpha\beta$-component of $\boldsymbol{\sigma}$, i.e., $\sigma_{\alpha\beta}$, is defined as the α-component of the force acting on a unit area element perpendicular to the β-axis.

The origin of stress in the Rouse model is the tension of the bonds connecting adjacent segments. (The stress of a polymeric liquid is composed of a contribution from the solvent and another from the polymer chains. In dense

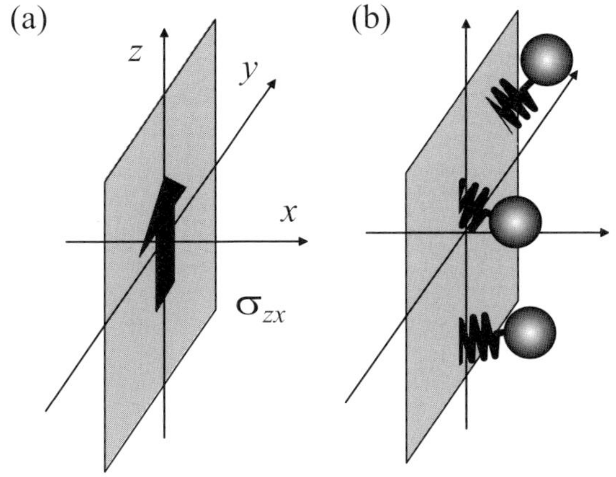

Fig. 5.12. (**a**) Definition of the stress tensor. σ_{zx} is the z-component of the force acting on a plane perpendicular to the x-axis. (**b**) The stress on a plane in a polymeric system originates from the springs that cross this plane

polymers, the latter contribution is dominant.) Let us consider a virtual plane P in a polymeric liquid. We assume that the normal to this plane is parallel to the β-axis. The force acting on the plane is caused by the coarse-grained bonds that cross it (see Fig. 5.12b). If we denote the i th coarse-grained bond vector ($i = 0, \ldots, Z-1$) by $\boldsymbol{u}_i$, the probability of finding this bond crossing the plane P is proportional to the projected length of $\boldsymbol{u}_i$ in the β-direction (the normal to P). The tension $\boldsymbol{u}_i$ in this bond is obtained by replacing the segment size b in (2.28) by the size of the coarse-grained segment a and differentiating it with respect to the end-to-end distance R. Therefore, the tension is given by $(3k_{\mathrm{B}}T/a^2)\boldsymbol{u}_i$. The contribution from this bond to the stress $\sigma_{\alpha\beta}$ is thus given by $(3k_{\mathrm{B}}T/a^2)(\boldsymbol{u}_i \cdot \hat{\boldsymbol{n}}_\alpha)$, where $\hat{\boldsymbol{n}}_\alpha$ is the unit vector in the α-direction. This contribution to $\sigma_{\alpha\beta}$ from the i th bond arises only when the i th bond crosses the plane P, and is proportional to $\boldsymbol{u}_i \cdot \hat{\boldsymbol{n}}_\beta$. Thus, summing the contributions from all bonds, we obtain

$$\sigma_{\alpha\beta} = n\frac{3k_{\mathrm{B}}T}{a^2}\sum_{i=0}^{Z-1}\left[\langle(\boldsymbol{u}_i)_\alpha(\boldsymbol{u}_i)_\beta\rangle - \frac{a^2}{3}\delta_{\alpha\beta}\right] , \tag{5.23}$$

where $(\boldsymbol{u}_i)_\alpha$ is the α-component of the bond vector $\boldsymbol{u}_i$, and n is the number of chains per unit volume. The second term on the right-hand side of (5.23) is the isotropic component of the stress, i.e., the hydrostatic pressure. It is easy to confirm that $\boldsymbol{\sigma}$ vanishes for the equilibrium state. To calculate the stress tensor using (5.23), we need the probability distribution of the bond vectors. Let us introduce the following bond-orientation tensor defined by

$$\boldsymbol{S}(i,t) \equiv \frac{3}{a^2}\langle\boldsymbol{u}_i(t)\boldsymbol{u}_i(t)\rangle - \mathbf{1} . \tag{5.24}$$

The second term on the right-hand side guarantees that this tensor vanishes when the bond distribution is in equilibrium. Combining (5.23) and (5.24), we finally obtain the following expression for the stress tensor:

$$\boldsymbol{\sigma}(t) = nk_{\mathrm{B}}T \sum_i \boldsymbol{S}(i,t) \ . \tag{5.25}$$

This relation shows that the stress relaxation of a polymeric liquid is caused by a relaxation process of the bond orientation distribution from the initial anisotropic distribution to the final isotropic equilibrium distribution.

Model for Stress Relaxation Based on Reptation Theory

Let us calculate the stress relaxation function $G(t)$ of an entangled dense polymer system based on the reptation theory. To do this, we consider a single Rouse chain in a tube and calculate the stress relaxation after a step strain is imposed on it at $t = 0$.

Definition of Tensors

Let us consider a spatially uniform linear transformation from a position vector $\boldsymbol{u}$ to a position vector $\boldsymbol{u}'$ by a transformation matrix $\boldsymbol{P} = (P_{\alpha\beta})$ as $\boldsymbol{u}' = \boldsymbol{P}\boldsymbol{u}$. Then, a second-rank tensor $T_{\alpha\beta}$ $(\alpha, \beta = x, y, z)$ is defined as a matrix that transforms by $\boldsymbol{P}$ as $\boldsymbol{T} \to \boldsymbol{T}'$ via the relation

$$T'_{\alpha\beta} = P_{\alpha\alpha'} P_{\beta\beta'} T_{\alpha'\beta'} \ ,$$

where each index that appears twice is summed over x, y, and z. (Strictly speaking, there are two types of index: one is the covariant one and the other is the contravariant one. However, this difference does not affect our discussions here.)

If the transformation matrix $\boldsymbol{P}$ is an orthogonal matrix, such a transformation can be decomposed into a rotation and (if necessary) a mirror reflection. An isotropic tensor is defined as a tensor that is invariant under any rotation (mirror symmetry is not required). In an isotropic system, any physical quantity that has tensorial nature is described by an isotropic tensor.

The first-rank isotropic tensor is the zero vector because of the isotropic symmetry. The 2nd, 3rd and 4th rank isotropic tensors are given by

$$C_{\alpha\beta} = C_1 \delta_{\alpha\beta} \ ,$$

$$C_{\alpha\beta\gamma} = C_1 \epsilon_{\alpha\beta\gamma} \ ,$$

$$C_{\alpha\beta\mu\nu} = C_1 \delta_{\alpha\beta}\delta_{\mu\nu} + C_2\left(\delta_{\alpha\mu}\delta_{\beta\nu} + \delta_{\alpha\nu}\delta_{\beta\mu}\right) + C_3\left(\delta_{\alpha\mu}\delta_{\beta\nu} - \delta_{\alpha\nu}\delta_{\beta\mu}\right) \ ,$$

where C_1, C_2 and C_3 are arbitrary scalar constants and $\epsilon_{\alpha\beta\gamma}$ is a 3rd rank tensor that takes the value $+1$ or -1 if $\{\alpha\beta\gamma\}$ is an even or odd permutation of $\{x, y, z\}$, respectively, and otherwise takes the value zero. This $\epsilon_{\alpha\beta\gamma}$ is usually called the Levi–Civita symbol or Eddington's epsilon symbol.

If the imposed strain is small enough, we can assume a linear viscoelastic behavior. In this case, the deformation of the chain is also small, and we can assume that the initial equilibrium conformation is almost preserved. Then the dominant factor in the dynamics of the chain in the tube is the Brownian motion of the chain due to thermal fluctuations. The diffusion constant of the one-dimensional diffusion of this Rouse chain in the tube is given by (2.99) as[3]

$$D_{\mathrm{c}} = \frac{k_{\mathrm{B}}T}{Z\zeta_{\mathrm{c}}}\,, \tag{5.26}$$

where ζ_{c} is the friction constant of a coarse-grained segment moving in the tube. We first consider the case where chains are very long compared to the tube diameter, i.e., $Z \gg 1$. The validity of this assumption will be discussed later. The central limiting theorem presented in Sect. 2.1 states that, when we consider a sum of N independent stochastic variables that obey the same probability distribution ($N \gg 1$), its standard deviation is smaller than its average by a factor of $1/\sqrt{N}$. Applying this theorem to the present polymer case, we find that the ratio of the fluctuation in the chain length to the total chain length is of the order of $1/\sqrt{Z}$, and is negligible in the long-chain limit. It is then reasonable to assume that the total chain length measured in units of a (the tube diameter) is kept constant with the equilibrium length (5.22). In such a situation, the motion of the chain is described by the diffusion of its center of mass.

Gaussian Distributions in Three Dimensions

Consider a three-dimensional Gaussian stochastic variable $\boldsymbol{u} \equiv (u_x, u_y, u_z)$. Its probability distribution function has the general form

$$P(\boldsymbol{u}) = \frac{1}{\sqrt{(2\pi)^3 \det \boldsymbol{C}}} \exp\left[-\frac{1}{2}(\boldsymbol{u}-\bar{\boldsymbol{u}})^{\mathrm{T}}\boldsymbol{C}^{-1}(\boldsymbol{u}-\bar{\boldsymbol{u}})\right]\,,$$

where $\boldsymbol{u}^{\mathrm{T}}$ is the transposed vector of $\boldsymbol{u}$, $\det \boldsymbol{C}$ is the determinant of the matrix $\boldsymbol{C}$, $\bar{\boldsymbol{u}} \equiv \langle \boldsymbol{u} \rangle$ is the mean value of $\boldsymbol{u}$, and $\boldsymbol{C}^{-1}$ is the inverse of the covariance matrix defined by

$$\boldsymbol{C} \equiv \Big\langle (\boldsymbol{u}-\bar{\boldsymbol{u}})(\boldsymbol{u}-\bar{\boldsymbol{u}}) \Big\rangle \quad \text{or} \quad C_{\alpha\beta} = \langle (u_\alpha - \bar{u}_\alpha)(u_\beta - \bar{u}_\beta) \rangle\,,$$

where $\alpha, \beta = x, y, z$.

As the covariance matrix is a real symmetric matrix by definition, it can be diagonalized by an appropriate orthogonal matrix $\boldsymbol{P}$ to give

$$\boldsymbol{P}^{\mathrm{T}}\boldsymbol{C}\boldsymbol{P} = \begin{pmatrix} \lambda_x & 0 & 0 \\ 0 & \lambda_y & 0 \\ 0 & 0 & \lambda_z \end{pmatrix}.$$

[3] While the discussion in Sect. 2.4.2 is for three-dimensional systems, the result is also true for one-dimensional systems with a modification of the coefficient.

Applying a linear transformation using this orthogonal matrix as $\boldsymbol{u}' = \boldsymbol{P}\boldsymbol{u}$, we can transform $\boldsymbol{u}$ to another stochastic vector $\boldsymbol{u}'$ whose components are statistically independent Gaussian variables. Then the variance of the α-component of $\boldsymbol{u}'$ is given by λ_α.

Due to the strain imposed at $t = 0$, the orientational distribution of each bond deviates from the isotropic one. This corresponds to the generation of shear stress according to (5.25), which relaxes with the relaxation of the deformed bond orientation. Because the chain is constrained in the tube, such a relaxation of the bond orientation can take place only at the chain ends where the chain can move freely (see Fig. 5.13). This situation corresponds to the following boundary condition at the chain ends:

$$\boldsymbol{S}(i,t) = \boldsymbol{0} \qquad (i = 0 \ \text{ or } \ i = Z - 1)\,. \tag{5.27}$$

On the other hand, in the middle of the chain, the bond orientation tensor diffuses along the chain due to the diffusion motion of the center of mass of the chain relative to the tube. Using the continuous description of the segment index ($i \to s$), the diffusion process is described by

$$\frac{\partial}{\partial t}\boldsymbol{S}(s,t) = \frac{D_\mathrm{c}}{a^2}\frac{\partial^2}{\partial s^2}\boldsymbol{S}(s,t)\,, \tag{5.28}$$

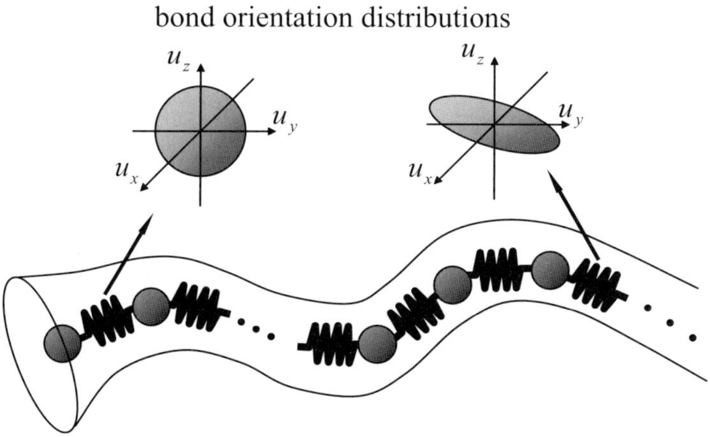

Fig. 5.13. Illustrative explanation of the concept of bond orientation tensor. The *upper two figures* show the distribution functions of the bonds at the chain end and at the center of the chain. At both ends of the chain, the bond orientation distribution is isotropic because there is no constraint due to the tube. If we assume that such a bond distribution is described by a three-dimensional Gaussian distribution, the deviation of its covariance matrix from an isotropic one corresponds to the bond orientation tensor (see the box entitled *Gaussian Distribution in Three Dimensions*)

where the factor a^2 on the right-hand side is the Jacobian, relating the index of the segment s to the distance along the chain.

Green's Function for Stress Relaxation

The stress relaxation function is obtained by solving the one-dimensional diffusion equation (5.28) with the boundary condition (5.27). Here, we consider the stress relaxation after imposing a step shear strain with $\gamma_0 = \mathrm{d}x/\mathrm{d}z$ instantaneously at time $t = 0$ (see Fig. 5.4b). In this case, among the 9 components of the orientation tensor $\boldsymbol{S}$, we need only consider the xz-component, S_{xz}, because the stress relaxation is fully described by this component.

As (5.28) is a linear partial differential equation, the solution starting from the initial condition $S_{xz}(s, t = 0) = \gamma_0$ is expressed as

$$S_{xz}(s,t) = \int_0^Z \mathrm{d}s' G(s',0;s,t) S_{xz}(s',0) \ , \tag{5.29}$$

where $G(s', 0; s, t)$ is the Green's function of the partial differential equation (5.28) under the boundary condition (5.27). The Green's function $G(s', 0; s, t)$ is defined as the solution of the equation

$$\frac{\partial}{\partial t} G(s',0;s,t) - \frac{D_\mathrm{c}}{a^2}\frac{\partial^2}{\partial s^2} G(s',0;s,t) = \delta(s-s')\delta(t) \ . \tag{5.30}$$

The explicit expression for the solution of this partial differential equation is given by

$$G(s',0;s,t) = \frac{2}{Z}\sum_{p=1}^{\infty} \sin\left(\frac{p\pi}{Z}s\right) \sin\left(\frac{p\pi}{Z}s'\right) \exp\left(-\frac{p^2 t}{\tau_\mathrm{d}}\right) \ . \tag{5.31}$$

By substituting (5.31) into (5.30), one can easily verify that (5.31) is the solution of (5.30). In (5.31), the parameter τ_d is defined by

$$\tau_\mathrm{d} = \frac{Z^2 a^2}{\pi^2 D_\mathrm{c}} = \frac{Z^3 a^2 \zeta_\mathrm{c}}{\pi^2 k_\mathrm{B} T} \ , \tag{5.32}$$

where the second equality is derived using (5.26). This τ_d is called the reptation time. Its physical meaning will be discussed in the next section.

Figure 5.14 shows how the stress relaxes according to (5.28). We can confirm that stress relaxation proceeds from the ends of the chain to the middle of the chain.

Relaxation Due to Reptation Motion

The reptation time is the characteristic time scale of the slowest relaxation mode, specified by $p = 1$, among the relaxation modes in (5.31). During this

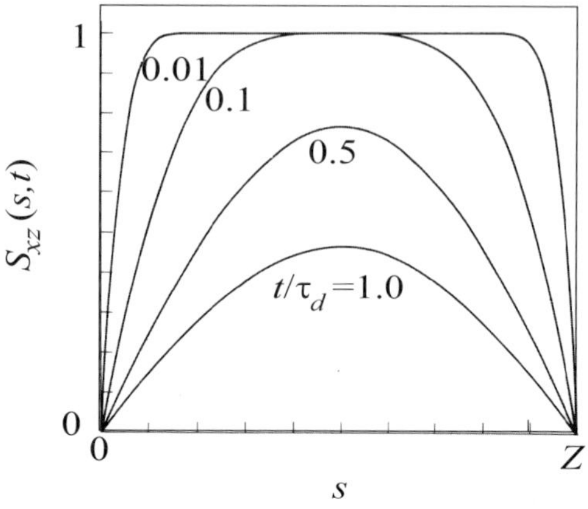

Fig. 5.14. Behavior of the relaxation of the anisotropy in the bond orientation obtained with the reptation theory. *From top to bottom*, the times are $t/\tau_\mathrm{d} = 0.01, 0.1, 0.5, 1.0$. Reprinted with permission from M. Doi and S.F. Edwards, *The Theory of Polymer Dynamics* (© 1986, Oxford Science)

reptation time, the chain diffuses one-dimensionally within the tube over its total length. Therefore, the chain escapes from the original tube after the reptation time. This escape process implies a relaxation of the topological constraint imposed by the tube, which describes the deformation of the network of entangled chains. Hence, the reptation time τ_d is a measure of the time scale of the stress and conformational relaxations due to the reptation motion. Equation (5.32) shows that the reptation time is proportional to Z^3. As the total number Z of coarse-grained segments is proportional to the total number of original segments N, we obtain the scaling relation:

$$\tau_\mathrm{d} \propto N^3 . \tag{5.33}$$

This relation indicates that the relaxation due to reptation is much slower than that of a Rouse chain given by (2.100), which is proportional to N^2. Because the reptation motion is essentially a one-dimensional diffusion motion of a Rouse chain confined in a tube, the reptation time given by (5.32) should be related to the Rouse relaxation time given by (2.100). Indeed, by relating $a \leftrightarrow b$, $Z\zeta_\mathrm{c} \leftrightarrow \zeta$ and $Z \leftrightarrow N$, these two expressions can be identified, except for a factor of 3 that originates from the difference in dimensionality. The origin of the different chain-length dependence of the relaxation times between the Rouse model and the reptation model [(2.100) and (5.32), respectively] is the extra factor N in the relation between the diffusion coefficient of the center of mass of the Rouse chain in the tube and the diffusion coefficient of one-dimensional diffusion of the individual segment.

As in the case of the Rouse model, one can obtain the molecular weight dependence of the viscosity for the reptation model. The elasticity of a dense polymer system originates from the deformation of the network formed by the chains. This elasticity depends on the characteristic size of the mesh of the network structure (given by the entanglement distance a) and does not depend strongly on the chain length N. On the other hand, the relaxation time of the reptation model obeys the scaling law $\tau_\mathrm{d} \sim N^3$ [see (5.33)]. Substituting these two scaling relations into (2.103), we obtain

$$\eta \sim G\tau \sim N^0 \times N^3 \sim N^3 \,. \tag{5.34}$$

Next, let us estimate the diffusion constant of the center of mass of the chain. After the time τ_d has passed, the entire chain escapes from the original tube. This means that the center of mass of the chain diffuses through a distance of the same order as the radius of gyration of the chain during the time τ_d. Hence, the diffusion constant D_G of the center of mass of the chain due to reptation motion is related to τ_d by $R_\mathrm{G}^2 \sim D_\mathrm{G}\tau_\mathrm{d}$, from which we obtain

$$D_\mathrm{G} \sim N^{-2} \,, \tag{5.35}$$

where we used $R_\mathrm{G} \sim N^{1/2}$, because a chain in a melt behaves as an ideal chain according to the screening effect.

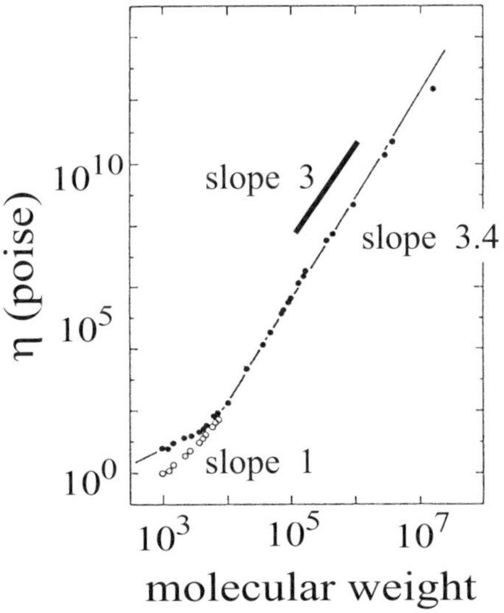

Fig. 5.15. Comparison of the molecular weight dependence of the viscosity between experimental data on a polybutadiene melt and theoretical results obtained using the reptation theory. In the reptation theory, the viscosity is estimated using the theoretical value of the longest relaxation time τ_d. For low-molecular-weight polymers, the viscosity η behaves as $\eta \sim N$, as predicted by the Rouse model in (2.101). For higher-molecular-weight polymers, the reptation theory predicts $\eta \sim N^3$, as given in (5.34), while the experimental data show $\eta \sim N^{3.4}$. Reprinted with permission from R.H. Colby et al., Macromolecules **20**, 2226 (1987). (© 1987, American Chemical Society)

It is known that the predictions of the reptation theory concerning the scaling properties of the relaxation time and the viscosity differ slightly from the experimental results. While (5.33) and (5.34) predict that both the longest relaxation time and the viscosity of a polymer melt are proportional to the cube of the molecular weight, the experimental data show a power law with the slightly larger exponent 3.4. Figure 5.15 shows the dependence of the experimental data for the viscosity on the molecular weight on a double logarithmic scale. This clearly indicates the 3.4 power of the molecular weight. In the next section, this discrepancy between theory and experiment will be explained by taking chain length fluctuations into account.

Viscous Relaxation and Diffusion of Polymer Melts

As shown in Fig. 5.15, the experimental data on the molecular weight dependence of the viscosity of a polymer melt show the dependence $\eta \sim N^{3.4}$, contradicting the prediction of the reptation theory, i.e., $\eta \sim N^3$. On the other hand, the experimental data on the diffusion constant D_{G} of polymer melts show the behavior $D \sim N^{-2}$ and this agrees with the prediction of the reptation theory given in (5.35). This different scaling behavior was not understood for a long time. Recently, the scaling behavior of the diffusion constant was re-examined using experimental data over a wide range of molecular weights. The experimental data show the scaling behavior $D_{\mathrm{G}} \sim N^{2.3}$, which is consistent with that of the viscosity (see Fig. 1.9a).

Stress Relaxation Function

Let us derive an explicit expression for the stress relaxation function using the reptation theory. Substituting (5.31) into (5.29), we obtain the following theoretical expression for the stress relaxation function $G(t)$:

$$S_{xz}(s,t) = \gamma_0 \sum_{p\ \mathrm{odd}} \frac{4}{p\pi} \sin\left(\frac{p\pi s}{Z}\right) \exp\left(-\frac{p^2 t}{\tau_{\mathrm{d}}}\right) \equiv \gamma_0 G(t) \,. \qquad (5.36)$$

This expression gives a theoretical prediction for the constitutive equation of the macroscopic viscoelastic behavior of polymer melts based on microscopic considerations using the chain model.

Figure 5.16 shows theoretical predictions concerning the storage modulus $G'(\omega)$ and loss modulus $G''(\omega)$, obtained from (5.36) and the corresponding experimental data for a melt of polybutadiene. The broken curves are the predictions of the reptation theory discussed in this section, whilst the continuous curves are those of the modified reptation theory where the effects of chain length fluctuations are taken into account. (The effects of chain length fluctuations will be discussed in the next section.) We see that the reptation theory reproduces the experimental data quantitatively.

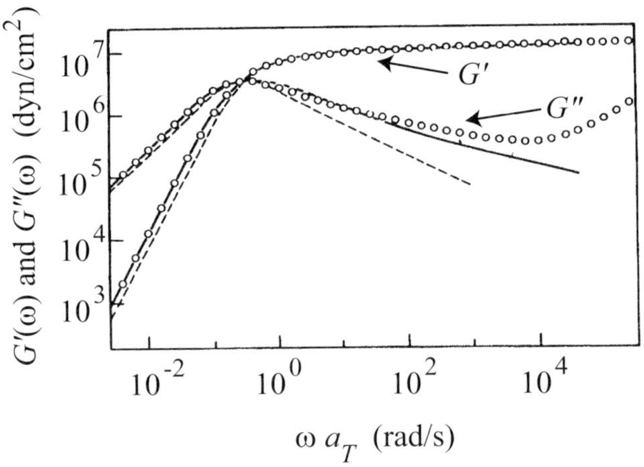

Fig. 5.16. Comparison between experimental data for the storage modulus $G'(\omega)$ and loss modulus $G''(\omega)$ of a polybutadiene melt, together with the corresponding theoretical predictions using (5.36), obtained by the reptation theory. Results of the original reptation theory (*broken curves*) and those of the modified reptation theory, taking the effects of chain length fluctuations into account (*continuous curves*) are shown. Reprinted with permission from D.S. Pearson, in *Rubber Chemical Technology*, Rubber Reviews (1987). (© 1987, American Chemical Society)

In the reptation theory, the stress relaxation is attributed to the gradual destruction of the tube due to the escape of the chain. Hence, the stress relaxation function $G(t)$ corresponds to the rate of the surviving part of the original tube after time t has passed. In this sense, $G(t)$ in (5.36) is sometimes called the survival probability of the tube.

5.3 Extensions of Reptation Theory and Nonlinear Viscoelasticity

When a dense polymer system is highly deformed due to a large external force, the linear relation given in (5.14) can no longer be assumed. In such a regime, the effects of an imposed strain γ_1 and another strain γ_2 are not superposed. Thus, the result of applying γ_2 after γ_1 is in general different from the result of applying γ_1 after γ_2. Such behavior is called nonlinear viscoelasticity. We now discuss the characteristic behavior of dense polymeric systems in the nonlinear viscoelastic regime.

Figure 5.17 shows experimental data for the stress relaxation function of a dense polystyrene solution on which a large instantaneous step strain is imposed at $t = 0$ [44]. The magnitude of the imposed initial strain increases from top to bottom. For the top curve, the imposed strain is so small that

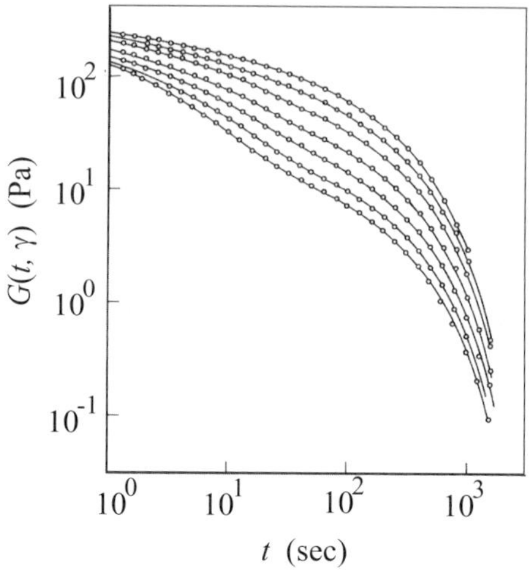

Fig. 5.17. Stress relaxation function of a dense solution of polystyrene on which a large step shear strain is imposed at $t = 0$. *From top to bottom*, the magnitude of the strain is increased. Reprinted with permission from K. Osaki et al., Macromolecules **15**, 1068 (1982). (© 1982, American Chemical Society)

the system is in the linear viscoelastic regime. The bottom curve is characteristic of the nonlinear viscoelastic regime. The main difference between the behavior in the linear and nonlinear regimes is the existence of fast relaxation in the early stage. (Note the shoulder in the short-time regime.) Apart from this short-time behavior, there is essentially no difference in the long-time behavior of linear and nonlinear viscoelasticity. The fast stress relaxation in the nonlinear regime suggests that there is another fast mode in the structural relaxation of the temporary network formed by the entangled polymers when it is highly deformed. Focusing on the chain dynamics in this nonlinear viscoelastic regime, let us modify the reptation theory discussed in the last section.

The reptation theory discussed in Sect. 5.2 was based on the picture of a chain with a constant length diffusing in a tube. However, this picture misses several important effects:

- contour length fluctuations,
- quick extension of the chain due to an external force and chain retraction,
- constraint release caused by disentanglement due to reptation motion.

Contour length fluctuations are important in explaining why the experimental reptation time depends on the chain length as $\tau_d \sim N^{3.4}$ in the linear viscoelastic regime. This contour length fluctuation is also important in the diffusion process of a branched polymer in an entangled polymer network. The second and third effects listed are pronounced in the nonlinear viscoelastic regime, where they lead to different relaxation behavior than in the linear viscoelastic regime. In the following, we will give intuitive explanations of these effects without going into the details.

5.3.1 Contour Length Fluctuations

When we introduced the reptation theory in Sect. 5.2, we assumed that the chain is long enough to ensure that the fluctuation in the chain length around the equilibrium value (5.22) can be neglected. In real situations, however, the total number Z of coarse-grained segments, measured in units of the number of segments between entanglement points N_e, is not so large. In this case, the fluctuations in the chain length due to thermal agitation can give a non-negligible contribution.

Figure 5.18 shows a schematic illustration of the effects of chain length fluctuations on reptation motion. For a chain with fixed length, the tube is destroyed at both ends due to the diffusive motion of the chain out of the tube. If the chain length is fluctuating, however, the distance for the chain to diffuse in order to escape from the tube is determined not by the average chain length but by the minimum chain length (see Fig. 5.18). Hence, the contour length fluctuation effectively decreases the total chain length, and leads to a decrease in the reptation time τ_d. When the chain length N increases, the effect of contour length fluctuation decreases as $1/\sqrt{N}$ due to the central limiting theorem. This means that the difference between the theoretical prediction of the reptation theory for the relaxation time τ_d and its actual value becomes large when the chain length becomes short. As shown schematically in Fig. 5.19, the extra drop in the relaxation time in the short-chain region gives a larger exponent than the prediction of the reptation theory ($\ln \tau_d \propto 3 \ln N$). This larger exponent asymptotically approaches the theoretical value 3 in the long-chain limit.

The effects of contour length fluctuations are particularly important in the diffusion of a branched polymer in a matrix of dense entangled polymers.

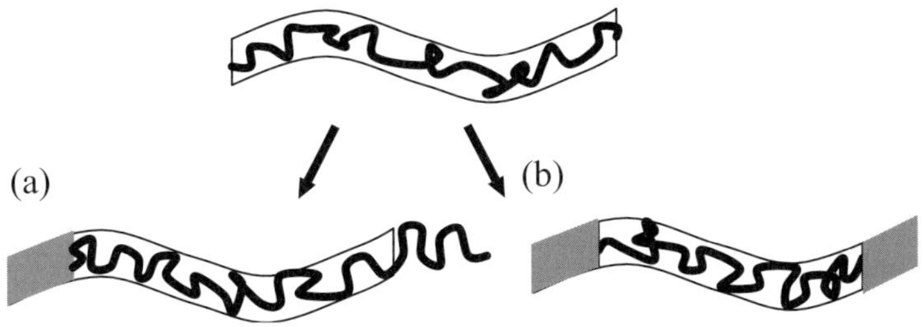

Fig. 5.18. (**a**) Disappearance of the tube due to diffusion of the center of mass of the chain (reptation) and (**b**) destruction of the tube due to the shrinkage of the thermally fluctuating chain. *Shaded regions* are the lost parts of the tube

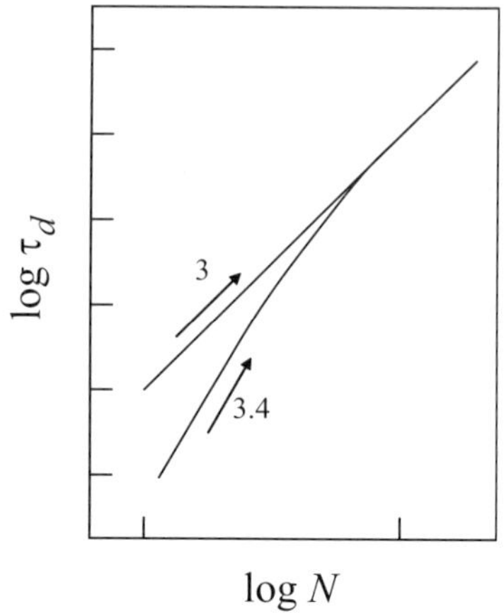

Fig. 5.19. Schematic illustration of the effective exponent of the power law dependence of the relaxation time on the chain length. The effective exponent is larger than the theoretical value 3 due to the effect of contour length fluctuations

Figure 5.20 is a schematic illustration of the diffusion process of a branched polymer in a dense polymer network. As one can imagine from this schematic illustration, the center-of-mass diffusion of a branched polymer is dominated by the diffusion of the branching point, which is called the junction point. In order for a junction point of a branched polymer to migrate, the branches connected to this junction point have to shrink below the average distance a between entanglement points. As such a shrinkage of the side chains is driven by thermal fluctuations, the probability for a branch composed of N_S seg-

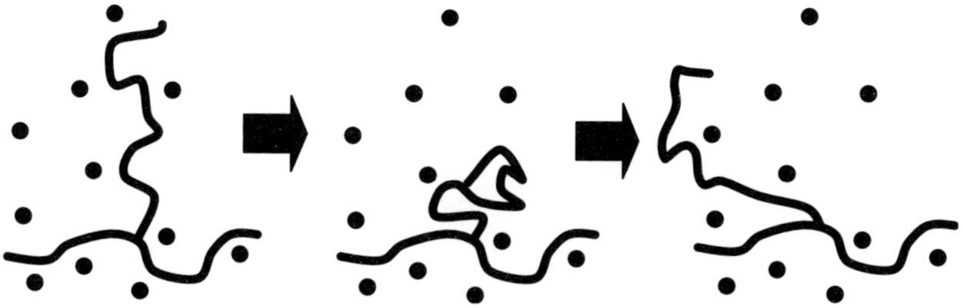

Fig. 5.20. Schematic illustration of the diffusion process of a branched polymer in a dense entangling polymer matrix. *Solid circles* show the obstacles due to matrix chains at which points the tagged chain cannot cross. The junction point can migrate only when a branch shrinks below the average distance between entanglement points

ments to shrink down to length l is obtained by assuming the conformational entropy of the ideal chain (2.26), viz.,

$$P(l) \sim \exp\left[-\frac{3k_{\mathrm{B}}T}{2N_{\mathrm{S}}b^2}(l-l_{\mathrm{eq}})^2\right] , \tag{5.37}$$

where l_{eq} is the equilibrium length of this branch, given using (5.22) as

$$l_{\mathrm{eq}} = \frac{b^2}{a}N_{\mathrm{S}} . \tag{5.38}$$

Using this probability distribution, we can calculate the probability for the branch to shrink down to length $a \sim 0$:

$$P(0) \sim \exp\left[-\frac{3k_{\mathrm{B}}T}{2}\left(\frac{b}{a}\right)^2 N_{\mathrm{S}}\right] . \tag{5.39}$$

This shows that the rate of the diffusion of the junction point is much slower than the reptation time of the individual branch. As a result, each branch contributes to the center-of-mass diffusion of a branched chain by a factor of $\exp(-\alpha N_{\mathrm{S}})$, where α is a constant independent of the length of the branch. The longer the branches (long-chain branching) or the greater the number of branches, the larger the suppression of the center-of-mass diffusion. Using this property, the viscosities of polymer materials are controlled by adjusting the numbers and lengths of side branches. Such a technique is important in polymer processing.

5.3.2 Chain Retraction

The experimental data of the stress relaxation shown in Fig. 5.17 indicate that some fast relaxation process other than reptation relaxation arises when a large strain is imposed. For simplicity, suppose that the temporary polymer network is affinely deformed when we impose an external uniform deformation. (The affine deformation corresponds to a uniform linear transformation of the coordinate system.) Under this assumption, the tagged chain and surrounding tube are stretched in the same manner. However, if the deformation is large, the highly stretched tagged chain rapidly shrinks in the tube to recover its equilibrium chain length. Such a motion corresponds to the relaxation of a stretched one-dimensional Rouse chain, whose characteristic time scale is the Rouse time $\tau_{\mathrm{R}} \sim N^2$ defined in (2.100). Note that this Rouse time is much shorter than the reptation time $\tau_{\mathrm{d}} \sim N^3$. Thus the relaxation associated with the rapid shrinkage of the chain inside the tube causes a rapid relaxation in the short-time regime. This is the origin of the shoulder in the short-time regime for the highly deformed case in Fig. 5.17.

5.3.3 Constraint Release

In the reptation theory discussed in Sect. 5.2, we assumed that the stress relaxation is caused by the destruction of the tube due to the escape of the chain from the tube. Recall that the concept of the tube is a mean field approximation of the entanglements between chains. Since an entanglement is formed by a pair of chains, when the tagged chain reptates out of the tube, it is not only the constraint on the tagged chain but also the constraint on the partner chain of the entanglement that is released (see Fig. 5.21 for a schematic illustration).

Such an effect is called constraint release or double reptation [45]. As this constraint release introduces an extra relaxation to the reptation theory, the stress relaxation becomes faster than was predicted in Sect. 5.2. It is known that this constraint release is important when describing the long-time viscoelastic properties of a mixture of long and short polymers. The stress relaxation on time scales longer than the reptation time of the short chains is dominated by entanglements among long chains. On such a time scale, short chains can be regarded as solvent molecules which do not contribute to the formation of the tube.

The mathematical formulation of constraint release is rather simple. In the original reptation theory given in Sect. 5.2, we assumed that the tagged chain diffuses amongst many static obstacles. This picture leads to the expression for the stress relaxation $G(t)$ given in (5.36). Recall that this $G(t)$ corresponds to the rate of the surviving part of the original tube after time t has passed. If we adopt the picture of the constraint release model, an entanglement point disappears when either of the two chains that form this entanglement reptates out. If we assume that the reptation motions of these two chains

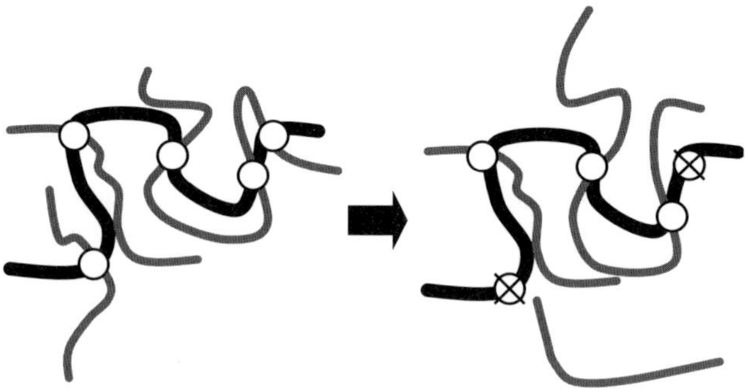

Fig. 5.21. Schematic illustration of the constraint release due to reptation motion. *Open circles* show entanglement points on the tagged chain (*bold curve*). Each entanglement point is formed by the tagged chain and the partner chain. The reptation motion of the partner chain contributes to the decrease in the number of entanglement points on the tagged chain

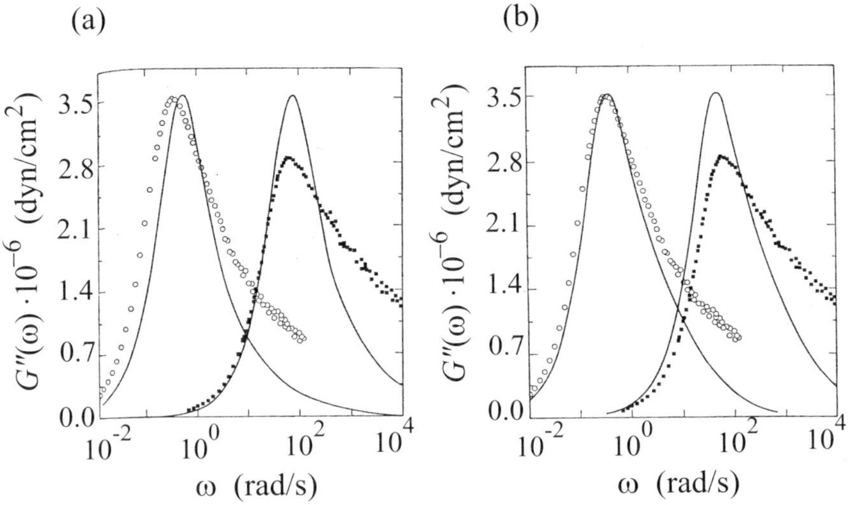

Fig. 5.22. Loss moduli $G''(\omega)$ calculated using (**a**) the original reptation theory (Sect. 5.2) and (**b**) the modified reptation theory, where the constraint release effect is included, together with experimental data for melts of monodisperse polybutadiene with molecular weights 3.55×10^5 (*left*) and 7.09×10^4 (*right*). [Note that in the original reference, (**a**) and (**b**) should be exchanged. This is due to a mistake in the printing process.] Reprinted with permission from J. des Cloizeaux, Europhys. Lett. **5**, 437 (1988). (© 1988, EDP Sciences)

are statistically independent, we find that the surviving probability of an entanglement point at time t is not $G(t)$ but $[G(t)]^2$. In Fig. 5.22, we show the predictions for the frequency-dependent loss modulus $G''(\omega)$ obtained using the original reptation theory and also using the reptation theory modified to include the constraint release effect. We also show the experimental data for polybutadiene, along with the theoretical predictions.

We can confirm that the agreement between theory and experiment is improved by introducing the constraint release effect for the case with higher-molecular-weight polybutadiene, for which the reptation theory is applicable. This indicates that the two-body effect is crucially important in the stress relaxation of an entangled polymer system, and that the single-body picture adopted in the original reptation theory is quantitatively unreliable.

There are several simulation studies based on the modified reptation theory that take most of the effects so far discussed into account [46, 47]. Figure 5.23 shows the stress relaxation function for a step strain obtained by numerically solving a set of Langevin equations for many bead–spring chains (see Sect. 2.4.1) which obey the modified reptation theory, including contour

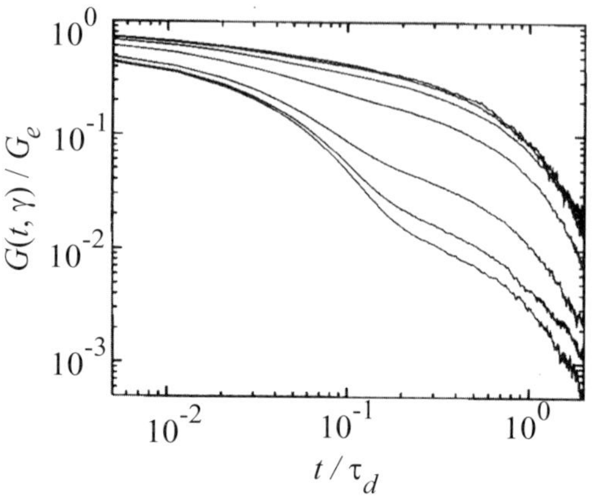

Fig. 5.23. Simulation results on the stress relaxation function of a polymer melt after a step train is imposed. The simulation was performed by solving a set of Langevin equations for many bead–spring chains obeying the modified reptation theory. The magnitude of the imposed strain increases *from top to bottom*. Reprinted with permission from C.C. Hua and J.D. Schieber, J. Chem. Phys. **109**, 10018 (1998). (© 1998, American Institute of Physics)

length fluctuations, constraint release and chain retraction. The simulation results reproduce the nonlinear viscoelastic properties shown in Fig. 5.17 quite well.

5.3.4 Contribution to Viscoelasticity from Phase-Separated Domains

There are some other important factors that contribute significantly to the viscoelasticity of polymer systems other than the reptation dynamics. One such example is the domain structures formed by phase separation. Figure 5.24 shows the temperature dependence of the behavior of the storage modulus $G'(\omega)$ around the transition temperature between the high-temperature disordered uniform phase and the low-temperature microphase-separated phase. It is obvious from this figure that the behavior in the high-frequency regime does not depend on the temperature and is essentially the same as for a single-component polymer melt. On the other hand, the storage modulus in the low-frequency regime is considerably increased when the microphase-separated lamellar domains are formed (the upper curves), i.e., the power law behavior of $G'(\omega)$ changes from $G'(\omega) \sim \omega^2$ (uniform phase) to $G'(\omega) \sim \omega^{0.5}$ (lamellar phase).

The behavior $G'(\omega) \sim \omega^2$ in the uniform phase can be explained using the slowest relaxation mode of the reptation dynamics in the long-time limit. On the time scale of the reptation time ($t > \tau_\mathrm{d}$), all relaxation modes with shorter characteristic time than τ_d decay, and the stress relaxation is described by a

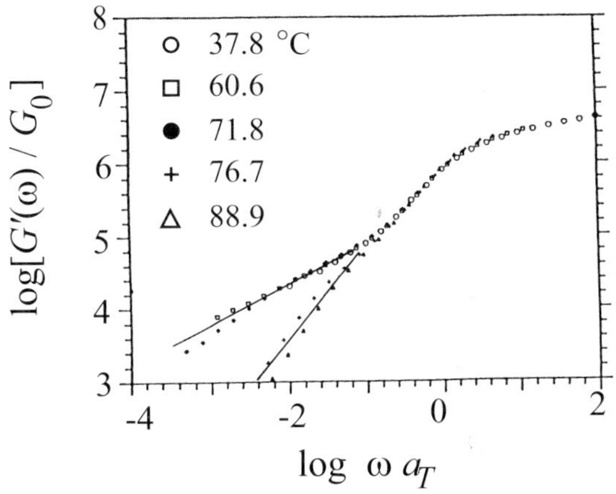

Fig. 5.24. Change in the storage modulus $G'(\omega)$ of a polyisoprene–polybutadiene block copolymer melt around its order–disorder transition temperature. The *lower curve* is for the high-temperature disordered uniform phase, and the *upper curve* is for the low-temperature microphase-separated lamellar phase. Reprinted with permission from M.D. Gehlsen et al., Macromolecules **25**, 939 (1992). (© 1992, American Chemical Society)

single exponential relaxation with the longest relaxation time τ_d:

$$G(t) = G_0 \exp\left(-t/\tau_\mathrm{d}\right) . \tag{5.40}$$

The elastic and loss moduli are then given as the real and imaginary parts of $i\omega\tilde{G}(\omega)$, respectively:

$$G'(\omega) = \frac{G_0 \tau_\mathrm{d}^2 \omega^2}{1+\tau_\mathrm{d}^2\omega^2} \sim \omega^2 , \qquad G''(\omega) = \frac{G_0 \tau_\mathrm{d}\omega}{1+\tau_\mathrm{d}^2\omega^2} \sim \omega \quad (\omega \ll 1/\tau_\mathrm{d}) , \tag{5.41}$$

which reproduces the power law behavior of the storage modulus $G'(\omega)$ in the low-frequency regime of the uniform phase. On the other hand, the frequency dependence of $G'(\omega)$ in the microphase-separated phase is attributed to the disordering of the orientations of the lamellar domains [48]. These results clearly demonstrate that a complete understanding of the macroscopic viscoelastic properties of inhomogeneous polymer systems requires knowledge not only of the microscopic dynamics of individual chains but also of the dynamics of mesoscopic structures, such as phase-separated domain structures. For such a purpose, a combination of the self-consistent field theory described in Chap. 3 and the reptation theory discussed in this chapter will be a suitable technique [49].

Exercises

Exercise 5.1. Imagine that an oscillatory strain

$$\gamma(t) = \gamma_0 \sin \omega t \; , \qquad \kappa(t) = \dot{\gamma}(t) = \gamma_0 \omega \cos \omega t \; , \tag{5.42}$$

is imposed on a viscoelastic body. Confirm that the real and imaginary parts of $\mathrm{i}\omega\tilde{G}(\omega)$ coincide with $G'(\omega)$ and $G''(\omega)$ in (5.4), respectively, where

$$\tilde{G}(\omega) = \int_0^\infty G(t) \mathrm{e}^{-\mathrm{i}\omega t} \mathrm{d}t \tag{5.43}$$

is the Fourier transformation of $G(t)$.

Exercise 5.2. Using the Rouse modes defined in Sect. 2.4.2, calculate the stress relaxation function $G(t)$ of the Rouse model on which a step strain is imposed.

References

1. P.G. de Gennes: *Scaling Concepts in Polymer Physics* (Cornell University Press, 1979)
2. P.M. Chaikin and T.C. Lubensky: *Principles of Condensed Matter Physics* (Cambridge University Press, 1995)
3. M. Daoud and C.E. Williams (Eds.): *Soft Matter Physics* (Springer-Verlag, 1995)
4. I.W. Hamley: *Introduction to Soft Matter* (Wiley, 2000)
5. G.R. Strobl: *The Physics of Polymers* (Springer-Verlag, 1996)
6. M. Doi and S.F. Edwards: *The Theory of Polymer Dynamics* (Oxford Science Publishing, Oxford, 1986)
7. R.G. Larson: *The Structure and Rheology of Complex Fluids* (Oxford University Press, New York, 1999)
8. A.S. Lodge: *Elastic Liquids* (Academic Press, London, 1964)
9. G. Gompper and M. Schick: In: *Phase Transitions and Critical Phenomena*, Vol. 16, ed. by C. Domb and J.L. Lebowitz (Academic Press, London, 1994)
10. J. Meunier et al. (Eds.): *Physics of Amphiphilic Layers* (Springer, 1987)
11. P.G. de Gennes and J. Prost: *The Physics of Liquid Crystals* (Oxford Science Publications, 1993)
12. H. Takayasu: *Fractals* (Asakura-Shoten, 1986) (in Japanese)
13. L.D. Landau and I.M. Lifshitz: *Statistical Physics* (Butterworth Heinemann, 1984)
14. J. Zinn-Justin, *Quantum Field Theory and Critical Phenomena* (Oxford Science Publishing, Oxford, 1989)
15. H. Yamakawa and G. Tanaka: J. Chem. Phys. **47**, 3991 (1967)
16. D.W. van Krevelen: *Properties of Polymers* (Elsevier, 1990)
17. N.G. van Kampen: *Stochastic Proccesses in Physics and Chemistry* (North-Holland, Amsterdam, 1981)
18. G. Ronca: J. Chem. Phys. **79**, 1031 (1983)
19. J.F. Marko and E.D. Siggia: Macromolecules **28**, 8759 (1995)
20. M.P. Allen and D.J. Tildesley: *Computer Simulation of Liquids* (Oxford Science Publishing, Oxford, 1987)
21. K. Binder (Ed.): *Monte Carlo and Molecular Dynamics Simulations in Polymer Science* (Oxford Science Publishing, Oxford, 1995)
22. R.P. Feynman and A.R. Hibbs: *Quantum Mechanics and Path Integrals* (McGraw-Hill, New York, 1965)
23. F.W. Wiegel: *Introduction to Path-Integral Methods in Physics and Polymer Science* (World Scientific, Singapore, 1986)
24. S.T. Milner, T.A. Witten and M.E. Cates: Macromolecules **21**, 2610 (1988)

25. K.M. Hong and J. Noolandi: Macromolecules **14**, 727 (1981)
26. G.J. Fleer, M.A. Cohen-Stuart, J.M.H.M. Scheutjens, T. Cosgrove and B. Vincent: *Polymers at Interfaces* (Chapman & Hall, London, 1993)
27. M.W. Matsen and M. Schick: Phys. Rev. Lett. **72**, 2660 (1994)
28. W.H. Press, S.A. Teukolsky, W.T. Vetterling and B.P. Flannery: *Numerical Recipes in C* (Cambridge University Press, 1992)
29. J.G.E.M. Fraaije: J. Chem. Phys. **99**, 9202 (1993)
30. A.V. Zvelindovsky, G.J.A. Sevink, B.A.C. van Blimmeren, N.M. Maurits and J.G.E.M. Fraaije: Phys. Rev. E **57**, R4879 (1998)
31. H. Morita, T. Kawakatsu and M. Doi: Macromolecules **34**, 8777 (2001)
32. D. Broseta, G.H. Fredrickson, E. Helfand and L. Leibler: Macromolecules **23**, 132 (1990)
33. E. Helfand and Z.R. Wasserman: In: *Developments in Block Copolymers*, Vol. I, ed. by I. Goodman (Elsevier, New York, 1982)
34. T. Ohta and K. Kawasaki: Macromolecules **19**, 2621 (1986)
35. H.S.M. Coxeter: *Introduction to Geometry* (John Wiley and Sons, New York, 1965)
36. U. Seifert: Adv. Phys. **46**, 13 (1997)
37. F.S. Bates, W.W. Maurer, P.M. Lipic, M.A. Hillmyer, K. Almdal, K. Mortensen, G.H. Fredrickson and T.P. Lodge: Phys. Rev. Lett. **79**, 849 (1997)
38. P. Hohenberg and W. Kohn: Phys. Rev. B **136**, 864 (1964)
39. L. Leibler: Macromolecules **13**, 1602 (1980)
40. G.H. Fredrickson and E. Helfand: J. Chem. Phys. **87**, 697 (1987)
41. T. Kawakatsu, R. Hasegawa and M. Doi: Int. J. Mod. Phys. C **10**, 1531 (1999)
42. S. Qi and Z.-G. Wang: Phys. Rev. E **55**, 1682 (1997)
43. S. Onogi, T. Masuda and K. Kitagawa: Macromolecules **3**, 109 (1970)
44. K. Osaki, K. Nishizawa and M. Kurata: Macromolecules **15**, 1068 (1982)
45. J. des Cloizeaux: Europhys. Lett. **5**, 437 (1988)
46. C.C. Hua and J.D. Shieber: J. Chem. Phys. **109**, 10018 (1998)
47. D.W. Mead, R.G. Larson and M. Doi: Macromolecules **31**, 7895 (1998)
48. K. Kawasaki and A. Onuki: Phys. Rev. A **42**, 3664 (1990)
49. T. Shima, H. Kuni, Y. Okabe, M. Doi, X.-F. Yuan and T. Kawakatsu: Macromolecules **36**, 9199 (2003)

Index

actin filament 7, 84, 90
action 109
adiabatic approximation 76
amphiphilic molecule 137, 149

backbone 9
bead–spring model 35
bending elastic energy 146
bending elasticity 84
biopolymers 90
block copolymers 12
block ratio 143
bond 18
bond orientation tensor 193
branched polymer 151
Brownian dynamics method 95
Brownian motion 62

central limiting theorem 32, 33
chain retraction 202
chain topology 19
Chapman–Kolmogorov relation 107
χ-parameter 60
classical limit 111, 112
coarse graining 81
coarse-grained model 34
coarse-grained segment 34
coefficient of viscosity 72
coil state 24
coil–globule transition 25
compatibilizers 149
compatibilizing effect 149
complex fluid 1
complex system 2
computer simulation 95
concentrated solution 55
configuration 12, 19
conformation 12, 19
constant mean curvature surface 147
constitutive equation 188
constraint release 202, 206
continuous phase transition 155
continuum limit 45
contour length fluctuation 202
copolymer 12, 19
correlation function 43
covariance matrix 195
creep measurement 182
critical
  micelle concentration 137
  phenomenon 154
  point 102
  temperature 155
cumulant 159, 160
  expansion 160
curvature 85, 86

Debye function 45
deformation velocity tensor 188
delta functional 129
density functional theory 158
depletion
  effect 115
  layer 115
die swell 181
diffusion 10
  constant 22
dilute solution 54
Dirac delta-function 30
discretization 137
DNA 7, 84, 90
double reptation 206
droplet 5
dyadic product 66, 108
dynamic elasticity measurement 182

effective bond length 36
Einstein's relation 65, 66
elastic body 180
elastic modulus 72, 182
elastic scattering 43
elasticity 72
emulsion 5
end-to-end distance 26
end-to-end vector 26
equation of continuity 186
excess free energy 158
excluded volume chain 21, 37, 49
excluded volume effect 21
excluded volume parameter 22, 49

FENE potential 96
Fick's law 141
filler 136
finite difference equation 138
first order phase transition 156
fixed point 51
Flory exponent 52
Flory–Huggins free energy 58
Flory–Huggins–de Gennes model 168, 169
fluctuating force 63
fluctuation–dissipation theorem 64
fluid particle 187
flux 64
Fokker–Planck equation 62, 64
form factor 139
Fourier transform 39
fractal curve 22
fractal structure 41
free draining 69
free energy 126
  landscape 177
full atomistic model 80
functional 122, 126
functional derivative 122, 124
functional group 9
functional integral 125
functional Taylor series expansion 125

gauche conformation 81
Gaussian chain 17, 40
Gaussian chain model 15
Gaussian curvature 146, 147
Gaussian integral 32
Gaussian stochastic variable 195
gel 5
Ginzburg–Landau free energy 162
Ginzburg–Landau theory 49, 153
globular state 12, 24
good solvent 24
grafting density 113
gyroid 144

hard core 23
Helfrich's bending energy model 146
homopolymer 12
Hookian elastic body 180
hydrodynamic interaction 74
hydrodynamically equivalent sphere 78

ideal chain 21, 37
incompressibility condition 74
inelastic scattering 43
interface 136
interfacial energy 145
interfacial tension 145, 171
intermediate scattering function 71, 73
isochronisms of pendulum 116
isotropic tensor 194
iteration method 135

junction point 204

kinetic coefficient 142
Koch curve 41
Kuhn length 82, 84

Lagrangian multiplier 134
Langevin equation 62, 63
lattice model 18
lattice random walk model 22
Legendre transformation 131
Lennard–Jones potential 23
light scattering 43
linear elastic body 180
linear response theory 162, 163
linear viscoelastic body 189
linear viscosity 75
linear viscous fluid 180
local equilibrium 134
local potential 169

loss modulus 72, 184

Maxwell model 189
mean curvature 146, 147
mean field 105
  approximation 37, 50, 103, 105
  theory 49
melt brush 115
mesoscopic scales 2, 8
micelle 137
microemulsion 5
microemulsion phase 151
microphase separation 102, 139, 142
microphase-separated structure 139
minimal surface 147
mixing entropy 57
molecular aggregate 136, 139
molecular dynamics method 95
molecular weight 19
monomer 9, 18
Monte Carlo method 95

natural coordinate 86
nearest neighbor lattice sites 18
neutron scattering 43
Newtonian fluid 180
non-draining 70
nonlinear viscoelasticity 189, 201
normal vector 86
number of segments between entanglement points 192

ordered bicontinuous double diamond 144
Oseen tensor 74, 76
overlapping density 54

path integral 105, 107
path integral representation 106
perforated lamellar 144
permutation 39
persistence length 82
perturbation theory 49, 53
phantom chain 70
phase separation 10, 101, 136
physical properties 10
polyelectrolyte 84
polymer 4
  brush 112, 136
  melt 55
  solution 54
polymerization reaction 9
poor solvent 24
pre-averaging approximation 78
primary structure 12
principal curvatures 146
projection operator 76, 77

quantum mechanics 110

radius of gyration 27
random coil 21
random coil state 12
random copolymers 12
random flight model 22
random phase approximation 48, 54, 162
random walk 21
realistic chain 21
renormalization group theory 50, 51
repeating unit 9, 18
reptation 97
  theory 72, 186, 190
  time 197
Rouse mode 67
Rouse model 67
Rouse time 70
rubber-elasticity plateau 185

scale free 40
scaling 27
scaling exponent of ideal chain 27
scaling property 40
scattering experiments 43
scattering function 47
Schrödinger equation 110
screen 55
screening effect 52, 55, 105
second order phase transition 155
second virial coefficient 24
segment 18
  interaction parameter 58
  volume fraction 56
self-avoiding random walk 21
self-consistent field 105, 106
self-consistent field theory 21, 49, 103, 106
self-similar pattern 41

semi-dilute solution 54
semi-flexible chain 84
shear stress 179
size of the coarse-grained segment 35
space–time correlation function 73
spinodal point 162
spontaneous curvature 146
square gradient energy 168
statistical independence 29
steepest descent method 130
step deformation 182
step strain 189
Stirling's formula 57
stochastic differential equation 63
Stokes approximation 75, 186
Stokes equation 75
storage modulus 72, 184
strain 180
stress relaxation
  function 189
  measurement 182
strong segregation 145, 153
structure factor 139
subchain 106
surface active effect 149
surfactant 149
survival probability of the tube 201
swelling 21
swollen brush 115

TDGL model 174
tensor 194
thermodynamic limit 32
θ-solvent 24
θ-temperature 25
time-dependent Ginzburg–Landau model 174
trans conformation 81
Trotter formula 109

unit tangent vector 86
united atom model 81

van der Waals force 23
velocity gradient tensor 187
viscoelastic body 71, 179, 180
viscoelasticity 1, 4, 179
viscosity 72
viscous fluid 179
Voigt model 190
volume force 75

weak segregation 153
Wick's Theorem 39
Wiener–Khinchin relation 47
worm-like chain 84

X-ray scattering 43

Zimm model 78

Printing: Saladruck, Berlin
Binding: Stein+Lehmann, Berlin